BREAKING THE BANKS

BREAKING THE BANKS

Representation and Realities in New England Fisheries, 1866–1966

MATTHEW MCKENZIE

University of Massachusetts Press
Amherst and Boston

Copyright © 2018 by University of Massachusetts Press
All rights reserved
Printed in the United States of America

ISBN 978-1-62534-391-8 (paper); 390-1 (hardcover)

Designed by Sally Nichols
Set in Adobe Minion Pro
Printed and bound by Maple Press, Inc.

Cover design by William Boardman
Cover art: Artist unknown; detail from comic book cover *Massachusetts Men Against the Sea*, courtesy of Massachusetts Seafood Council (1967).

Library of Congress Cataloging-in-Publication Data

Names: McKenzie, Matthew G., author.
Title: Breaking the banks : representations and realities in New England fisheries, 1866–1966 / Matthew McKenzie.
Description: Amherst : University of Massachusetts Press, [2018] | Series: Environmental history of the northeast | Includes bibliographical references and index. |
Identifiers: LCCN 2018019183 (print) | LCCN 2018024507 (ebook) | ISBN 9781613766385 (e-book) | ISBN 9781613766392 (e-book) | ISBN 9781625343918 (paperback) | ISBN 9781625343901 (hardcover)
Subjects: LCSH: Fisheries—New England—History—19th century. | Fisheries—New England—History—20th century. | Haddock fisheries—New England—History.
Classification: LCC SH221.5.N4 (ebook) | LCC SH221.5.N4 M35 2018 (print) | DDC 639.20974—dc23
LC record available at https://lccn.loc.gov/2018019183

British Library Cataloguing-in-Publication Data
A catalog record for this book is available from the British Library.

Portions of chapter 2 were published in a previous form in "Iconic Fishermen and the Fates of New England Fisheries Regulations, 1883–1912" in *Environmental History* 17 (January 2012): 3–28; used by permission of Oxford University Press on behalf of the American Society for Environmental History and the Forest History Society.

For Merry, who showed me how to bring good into the world
For Dave, who taught me that sometimes you have to fight for it
And for Shannon and Sam, as always, for giving me a reason to do so

CONTENTS

Preface ix

Introduction 1

CHAPTER 1
MOBILIZING A TRADITION
Gloucester and the Demise of the Schooner Fishery, 1866–1892 13

CHAPTER 2
THE BENEFITS OF MODERN FISHING
Boston and the Birth of Heavy Trawling, 1905–1925 36

CHAPTER 3
MASKING INDUSTRIAL REALITIES
Schooner Racing, Industrial Fishing, and Federal Aid, 1919–1939 61

CHAPTER 4
REINVENTING TRADITION
New England Embraces Industrial Fishing 95

CHAPTER 5
THE CONSEQUENCES OF MODERN FISHING
Stock Declines, Labor Unrest, and International Competition 117

CHAPTER 6
INVOKING THE PAST, IGNORING THE PRESENT, COMPROMISING THE FUTURE 137

Conclusion 168
Notes 185
Index 201

PREFACE

One morning in 1986, as my school bus ran its normal route through Woods Hole, Massachusetts, I looked out the window to see something I had never seen along the route before. As the bus chugged past the fishing pier, I saw dozens of fishing vessels tied up alongside one another where before I had rarely seen more than a few. That night, I asked my dad if he knew what was going on. "There's no fish left on Georges," I recall him replying. As the weeks passed, I noticed a palpable edge among many of the people my dad worked with in town. He was a contractor—not a fisherman—but some of his fellow contractors served the town's small fishing fleet. For this sixteen-year-old, it was thoroughly unsettling to see fear in the eyes of people not accustomed to fearing much of anything.

Driving all of this was a crash in the Georges Bank cod landings. From the usual landings of roughly 220 million pounds a year (100,000 mt), fishermen in 1986 only found about half that amount. Many openly talked, clearly concerned, about the end of New England groundfishing. The scientists in Woods Hole also voiced concerns, and as my family was friends with both scientists and families with ties to the fishery, we saw the full range of worry.

Shortly after the cod crash (little did we know then how much farther landings would fall), many of the town's few fishing vessels disappeared. The dock I saw from the bus every morning grew increasingly still and empty. During the summers, I bussed tables and mowed lawns to pay for college, watching dock activity grind down to nothing. By the time I left for college, I rarely saw fishing vessels there anymore. We had helped a few of my dad's fellow contractors move away as their search for work took them off Cape Cod. Although I didn't see the connection at the time, I later realized that my work in the tourist industry was intimately linked to the decline of the fishing industry. As one moved out, unable to sustain itself on the declining groundfish stocks, the other moved in.

Now, some thirty years later, I find that fearing the end of New England fishing has defined my entire adult life. Federal agencies routinely declare the groundfish fishery to be a disaster, and the fleet has been hemorrhaging active participants for years. Fishermen, researchers, and regulators have tried for decades to turn things around, to no avail. And while the public, when it pays attention at all, places blame on one or all of those groups, I realized as all this unfolded that we—the New England and even the American public—have never considered what role we had in fomenting and propagating the decline of our region's marine resources. Through the pages that follow, I hope readers will see the important role the wider public plays in marine resource use and management. It's not just about numbers: it's about context, and for too long we've ignored that context.

In this book, I try to resolve an apparent paradox: why, in a region that defines itself more by groundfish fishing than almost any other, have we utterly failed to effectively sustain those culturally iconic groundfish stocks? Throughout the following pages, I hope readers will come to understand that when we confuse romance with industrial reality, not only do we ignore our own resource base but we harm the very people who harvest those resources for our benefit. In a nutshell, between 1866 and 1966, New Englanders held onto a vision of our fisheries that was becoming increasingly separated from their industrial realities. Where we saw timeless continuity embodied in New England's fishing tradition, the industry was, and had always been, undergoing dynamic and tumultuous change. Where we celebrated our fishing heritage, that heritage itself hid how much the industry had already lost. How and why that happened forms the subject of this study.

This book owes much to the many people and institutions with whom I have had the privilege to work during the past seven years. First and foremost I must thank my fellow historians who offered yeoman service in listening to ideas and slogging through rough manuscript copy: Brian Payne, Matthew Booker, and Colin Davis all helped make this project tighter and more focused. At the University of Connecticut's Avery Point campus, Noreen Blaschik proved a godsend in formatting figures and catching manuscript errors. Both Brian Halley and Rachael Hoy DeShano with the University of Massachusetts Press proved ardent, constant, and steady supporters of the project, and were an absolute pleasure to work with. Most importantly, however, this project allowed me to work again with one of the most insightful editors working in marine environmental history today, Karen Alexander. Few know the science and history as well as she does, and her willingness to work with me on the manuscript was truly fortunate. I have been blessed by her friendship for over two decades, and in this book, further blessed by her dedicated, sometimes even stubborn, persistence and her (almost) always correct editorial instincts.

Colleagues from the Northeast and Atlantic Region Environmental History Forum (NEAREH) patiently considered dead fish more than any human being should: again, Brian Payne, but also Ed MacDonald, Claire Campbell, and Tony Penna. These dedicated souls and ideal colleagues lent an ear—willingly or no—to help me wrestle through the tangles that historians know so well. My fellow Avery Point colleague Michele Baggio lent his expertise to help me work through the economic complexities that may, or may not, have influenced this study. Similarly, Manuel Lizarralde, Juha Sarimaa, and John Lefeber all patiently listened to dead fish stories by campfires and pulled me into the woods to hunt deer when I needed to get out. I didn't get many, but not due to lack of their tutelage and support.

This project grew out of an emerging community of interdisciplinary historians and marine ecologists stitching together historical marine ecology and marine environmental history for almost twenty years now. Discussions and collaborations with Emily Klein, Ruth Thurstan, Loren McClenachan, Bill Leavenworth, and Adrian Jordaan proved, frankly, inspirational. Many of these people also provided their insights when this work was presented at the September 2017 International Coalition for the Exploration of the Seas' (ICES) meeting of the History Working

Group (WGHIST) in Lysekil, Sweden, funded through the University of Connecticut's Research Foundation and through History Department research funds.

While writing this book, I have enjoyed the privilege of representing Connecticut on the New England Fishery Management Council, the body tasked with developing commercial fisheries regulations in the region's federal waters out to two hundred nautical miles. Within that community, I worked with remarkably helpful people, and while we did not always agree on specific issues that the council deliberated, their insights helped me see this study in a completely different light. It is not too much to say that without that experience, and without working with these people, I could not have made any sense of New England's industrialized and mechanized fisheries. Erica Fuller suffered through interminable blathering about historical fishing, as did Jud Crawford, Aaron Dority, John Pappalardo, Fiona Hogan, Michelle Bachmann, Melanie Griffin, and Jamie Cournane. I also benefited from discussions with Gib Brogan, Kasia Deuel, Greg Wells, Aaron Kornbluth, and Morgan Callahan. Through council service, I was also able to continue my apprenticeship in ecology with my Avery Point colleague Peter Auster, and Mike Fogarty gave me a crash course in ecosystem modeling that lined up many seemingly disparate ecological dots. I also enjoyed the privilege of working with and getting to know Ted Ames, whose ecological and historical insights offer the best promise for getting New England's fisheries back onto the right track.

I'm also grateful for the time that Terry Alexander, Peter Kendall, Vincent Balzano, Terry Stockwell, Eric Reid, John Quinn, and John Bullard spent educating me on the operational and regulatory issues that make New England fishing such a puzzle. This is the second time I have had the pleasure to work with John Bullard, and for that I feel greatly fortunate. Finally, and most importantly, the council allowed me to work with, and become friends with, Dave and Merry Preble. Words can't express the gift that Dave and Merry's friendship has meant to me, my family, and this book. Merry's strength and dogged goodness simply stuns and inspires. Dave challenged me to embrace my potential. Through our conversations about politics, biology, history, writing, and nature, and his continued dedication to fighting for good in this world, he remains a rare blessing that continues to profoundly enrich my life.

A University of Connecticut Humanities Institute Fellowship in 2012 gave me the time at an early stage to lay down a solid foundation for the rest of the project. Glenn Gordinier and Eric Roorda provided me with several opportunities over several summers to present preliminary findings at the annual Munson Institute for Maritime History at Mystic Seaport. Mystic Seaport helped me stitch things together too, by providing me a "field research" berth aboard their restored 1949 eastern-rig dragger, F/V *Roann*, during an overnight transit from Woods Hole, Massachusetts, to Mystic, Connecticut. I must also thank the Massachusetts Historical Society's Boston Area Environmental History Seminar for hearing early drafts of this thesis and for providing an important venue for this kind of work for over twenty years. Oxford University Press has also generously granted permission to reprint some material previously published in *Environmental History* in 2012. Finally, I must thank the University of Connecticut Research Foundation for funding presentations at the American Society for Environmental History meetings in Chicago and Madison, Wisconsin, and at the History of Science Society meeting, also in Chicago.

Finally, I wish to thank, again, the magnificent people who supported and shepherded this project along at the University of Massachusetts Press. Dick Judd supported the project in its infancy, as did Tony Penna, who was forced to listen my thinking while captive in a six-hour drive to Bucknell. Brian Halley was equally supportive, and the editorial work of Rachael Hoy DeShano and Margaret Hogan was the best any author could ask for.

And, as always, none of this would have been possible without Shannon and Sam. If this work is any good, it is because of the people I mentioned above. If it's not, well, that's my fault.

BREAKING THE BANKS

INTRODUCTION

In December 2013, I sat in a hotel conference room representing Connecticut, my home state, on the New England Fishery Management Council. Before us lay a truly horrible decision: should the council ignore the recent Gulf of Maine cod stock assessments and allow unsustainable levels of fishing to remain in place, or, following mandates laid out in the Magnuson Fisheries Conservation and Management Act, reauthorized in 2007, should we abide by the scientific findings and slash groundfish quotas by up to 75 percent to achieve sustainable catch levels—and wreck fishermen's livelihoods?

These were lousy options. For weeks, the *Boston Globe* had run stories detailing the dire consequences that would unfold for individuals, families, and communities if quotas were cut. Often exploring the issue through an individual fisherman's circumstances, these stories forecast economic ruin and disrupted regional traditions if fishermen lost access to local marine resources. Other media outlets—local talk shows, radio, and newspapers—presented similar coverage. By the time the council met to deliberate, fishermen and their representatives had filled the room with expectations and apprehension. To begin the meeting, we grilled the scientists who ran the stock assessments. Then we heard industry lobbyists present well-considered and informed critiques of the scientific findings, and digested what lawyers from the National Marine Fisheries Service (NMFS) told us we could, and could not, legally do. Finally, as this was a public process, council chairman Terry Stockwell opened the meeting to public comments. Soon, the line of people wishing to comment on the public record stretched across the conference hall.

As they spoke, a few trends became clear. Some supported the cuts, resigning themselves to the sad reality of how depleted New England groundfish stocks had become. To these individuals, the only responsible way forward was to give the ecosystem a chance to recover. The vast majority of people, however, stridently opposed cutting quotas. Over a span of three hours, they pleaded, berated, threatened, and bullied council members to vote against the cuts. Nor were these public comments made at random. As the morning progressed, they took on an organized and orchestrated dynamic. The most respectable and reasonable industry representatives spoke first, presenting alternatives designed to work within current structures to find a better way forward. They were followed by a series of speakers who intensified the rhetoric, laying out the economic consequences and, ultimately, the cultural costs incurred if the groundfish fishery failed. Some of the most impassioned returned to their seats once they had finished speaking and started joking with their neighbors—as if the whole thing was a lark. Most, however, took the affair extremely seriously. They spoke clearly, passionately, and there was little doubt of their sincerity.

As a target of this choreographed presentation, I can heartily say that it was impressive and effective. I had come to the meeting resigned to vote in favor of the cuts as the only ecologically responsible choice. In the end, however, battered by vehement public testimony, the council postponed the vote for a couple of months. That bought us time to rerun the stock assessment using different industry recommendations and to explore options that might forestall slashing groundfish quotas. After three hours of increasingly acrimonious testimony, I leapt greedily at the chance to put off our decision, work up new ideas in hopes of finding a reasonable solution, and go home.

And later, I realized, *that* was the point. Weeks before the meeting, all interested parties prepared to use whatever arguments they could muster to get their desired results. Industry lobbyists, often lawyers, orchestrated public comments to delay or defeat quota cuts; environmental nongovernmental organizations (NGOs), often other lawyers, did the same to bolster the ecological case for imposing cuts; the NMFS and the Northeast Fisheries Science Center presented positions aiming ultimately to justify the policy that validated their models. Each interest group had the opportunity to make its case. This was the explicit intention of the Magnuson Act, which laid out the federal system for marine fisheries management, and of the National Environmental Policy Act, which laid out the mechanism for public input into federal environmental policymaking. On the surface, the issue appeared to pit economics versus ecosystem. Beneath that veneer, however, lay a complex array of competing political campaigns, each seeking to bring the council to one end or another. Industry representatives, NMFS scientists and decision-makers, environmental NGOs—their jobs were to bump the public process in their constituents' desired directions. To do that, groundfish industry lobbyists used all the tools available to them, be they sophisticated critiques of spawning stock biomass model construction or impassioned appeals to heritage and tradition. Without judgment, skepticism, or cynicism, I came to see that day the important role lobbyists played in the entire proceedings. They did nothing wrong or even questionable. They were doing their jobs, and on that day, they did them well.

In the end, however, the delay gave us nothing. Rerunning the stock assessment models showed that the stock was in worse shape and experiencing heavier fishing than we had thought. Three months later, we slashed the quotas.

I knew well what would come from this vote. Leaner quotas would lead to smaller landings, which undermine fishermen's ability to operate, and send economic ripples throughout their communities. Yet, many in that audience also feared the other, less quantifiable costs that would come from quota cuts. Not only were fishing businesses in danger of going under; these cuts also threatened to end a four-hundred-year multigenerational industry that had supported, according to one testifier at the December meeting, his ancestor's service against the British in the American Revolution. Cutting quotas threatened the continuance of a venerable pillar of American culture. Such comments made clear that people both inside and outside the industry saw New England fishermen as more than just businessmen: they were cultural icons protecting and preserving America's founding traditions and values. For these people, New England fishermen should be allowed to keep fishing in order to maintain living linkages to American heritage, whether fisheries were ecologically sustainably or not.

What became manifestly clear in our debates about cutting quotas, however, was how little any of us, myself included, knew about the recent history of New England's fisheries. As a historian serving as a federal marine fisheries manager, I was struck by how many comments invoked the past, but also by how often they rested on conventional wisdom and regional mythology rather than on empirical historical evidence. As testifiers claimed, New England has been home to fishing for over four centuries. None of us at the council table refuted these claims. But historical inconsistencies come into play. Contemporary groundfishing technologies—relying on satellite navigation, water temperature sensors, real-time satellite communications, powerful marine diesel engines, and increasingly sophisticated otter trawls—have very little in common with the way fishermen earned their living before 1925. Nor were current fishing operations necessarily part of the small-scale, family-owned fisheries that New England's lore has long claimed represented the industry. In short, New England's fisheries history seemed to stop with the end of the schooner fleet. In assessing how to integrate claims of heritage and tradition within a modern, scientifically driven fisheries management system, we were all basically winging it.

The following pages seek to fill that historical void. I argue that divergence between the cultural representation and industrial reality of these fisheries undermined meaningful attempts at fisheries regulation between 1866 and 1966. A number of significant findings emerge. First, this study shows that, in terms of capital organization, labor relations, politicization, and industrial dynamics, New England's groundfish schooner fishery was industrialized long before it was mechanized.

It also rescues from historical oblivion Boston's heavy haddock trawler fleet, its rise and fall, the thousands of unionized workers employed in it, the hired lobbyists working for both labor and management who represented the fleet's interests in Washington, and the political influence it wielded over federal fisheries discussions before the mid-1960s.

Developing in earnest in the mid-1920s, otter trawling, especially the largest vessels targeting haddock, took only three decades to seriously undermine the viability of local stocks of commercial species.

Most importantly, I argue that, while the fleet industrialized, mechanized, and then systematically jeopardized New England's fish stocks, New Englanders and Americans in general continued to hold romantic images of the region's preindustrial fishing past when they thought about fisheries at all—despite clear evidence to the contrary. For industry representatives and politicians, those images provided ready-made justifications for eschewing sustainable fisheries regulations—even when the industries themselves sought them. As the heavy trawler fishery fought for its life in Washington, D.C., in the 1950s and 1960s, those images provided all parties with convenient, publicly acceptable excuses for failing to restrain fishing pressure, even though it was abundantly clear that New England stocks could no longer sustain industrial scale extraction.

In tracing the industrial, cultural, and political history of New England's groundfish fisheries between 1866 and 1966, this study departs from the usual approaches to New England's marine environmental history. Previous academic studies, including most recently W. Jeffery Bolster's *The Mortal Sea* (2012), focused on the romantic sail-and-oar, hook-and-line fisheries and treated fishing in isolation from larger American society.[1] His and other analyses end with the mechanization of fishing in the early twentieth century as steam and diesel trawlers replaced the far more visually appealing schooner fleet. There is nothing wrong in celebrating the romantic in a compelling human endeavor. However, this enduring perspective obscures important historical trends—capital concentration, marginalized labor, ecological degradation—that carry through the transition from sail to steam.

Other previous studies, such as Margaret Dewar's *Industry in Trouble* (1983) and Peter B. Doeringer, Phillip I. Moss, and David G. Terkla's, *The New England Fishing Economy* (1986), paid scant attention to long-term historical trends by offering more pointillist twentieth-century analyses.[2] It would have been hard enough to explain the complexity of contemporary fishing without extending such in-depth studies into the past. Although detailed and insightful for contemporary policy questions, little connective tissue links these studies with others along a broader time span. Trade writers followed suit. Mark Kurlansky's *Cod* (1997) and *The Last Fish Tale* (2009), David Dobbs's *The Great Gulf* (2000), Paul Greenberg's *Four Fish* (2011), and Richard Adams Carey's *Against the Tide* (2000) all offer compelling stories, but their attention to contemporary fisheries issues is episodic. None offers a long-term, empirical, systematic historical analysis that frames current events within their larger economic, ecological, cultural, scientific, and political trajectories, which originated in the late 1800s but shaped the twentieth century.[3]

This work challenges from the outset a number of widely held assumptions about New England's fisheries that have hampered historical study. The first assumption is that we can discuss New England fishermen as workers in a singular industry—fishing—or as workers comprising a collective identity—fishermen. We can't. There has never been a prototypical New England "fisherman." New England's fisheries have long been split along multiple, shifting lines defined by gear, ethnicity, homeport, target species, and degree of capitalization. New England fishermen fractured along these issues long before mechanization, and continue to be so divided today.

My book explores one subset of this diverse and conflicted industry—the groundfish fishery from 1866 to 1966, which targeted flounders, haddock, redfish, and cod, among other species. It existed in parallel with many other fisheries—for instance, the swordfish, mackerel, herring, lobster, and scallop industries. Individuals often participated in several fisheries, however, each operating independently, as the differing behavior, biology, processing, and markets of target species shaped how people sought, caught, and harvested the stocks. Within the groundfish fishery, I focus in particular on the rise and fall of Boston's haddock fleet. While the mackerel fishery makes an appearance early on, and scalloping appears near the end, their stories deserve separate treatments.

The second assumption that needs challenging is that fisheries are insular, existing outside and independent of larger cultural frameworks. They are not. Fishing has always existed within a larger social context that frames, supports, and interprets how fishing, fishermen, and management are viewed and validated, or repudiated. Previous studies have embraced fisheries as a cultural activity integral to New England identity as a whole. Therefore, we must envision this set of practices in dialogue with the larger societal trends that shape it. Canadian researchers Ian McKay and Miriam Wright have analyzed such intersections in depth.[4] This monograph builds on their critical studies of the modern and antimodern within coastal communities and fisheries. Like their work, it seeks to situate fishing within the larger contexts of fisheries' industrializing, modernizing, and expanding beyond their small-scale, undercapitalized roots, at precisely the same time when many Americans were looking to fishing communities for reassurance and solace against those same dehumanizing forces. Not surprisingly, throughout the nineteenth and twentieth centuries, both of these threads created political tools that allowed some industry lobbyists to play politics better than others.

Thus, this book is not so much about fisheries or even fishermen. Rather, it explores a cultural vision of New England fisheries that had very little to do with the fishermen themselves. Relatively few working fishermen appear in these pages, largely for two reasons. First, few fishermen spoke out on these issues, deferring instead to hired spokesmen working for their unions or the vessel owners. Once these representatives may have worked on vessels, but at some point they ceased to be fishermen and became lobbyists or business managers. As a result, it was the people speaking on behalf of the fisheries, not the actual fishermen themselves, who constructed this powerful cultural representation to serve the ends of their constituents.

Second, over the century under consideration, conversations about fishing have not included much about actual fishermen. Instead, they present caricatures of working fishermen, often created outside the fishery. At no discernable point did anyone ask actual fishermen whether they *wanted* to be seen as icons of days gone by, or germs of an Anglo-American nation, or virtuous and docile laborers within a capitalist system. Indeed, it has become traditional for New Englanders to impose upon fishermen motivations, values, characteristics, and sentiments utterly divorced from those embraced by, embodied, agreed with, or reflected by the fishermen themselves. Therefore, in these pages I examine how cultural attitudes toward an idealized fishery often shaped fisheries regulation more powerfully than actual industrial experiences.

Since Roderick Nash published *Wilderness and the American Mind* (1967), environmental historians have identified the sentimentalization of nature as a critical cultural turning point in the conservation of American wilderness areas. Writers, photographers, painters, and filmmakers used imagery to create a love for wild areas that, in many cases, marginalized—even criminalized—subsistence workers extracting natural resources from them.[5] In rediscovering the fate of New England's groundfish industries, large and small, this study challenges environmental historians to consider critically the effects of sentimentalization on American wilderness management and use. As New England romanticized its historical ties to the sea, fishermen faced no such marginalization or criminalization (although their union eventually was broken for price fixing). Rather, New England's sentimentalization of the sea involved the celebration—perhaps the apotheosis—of fishermen harvesting local waters. Business-smart and politically savvy, New England's fisheries' leaders wasted no time in turning cultural affinity into policy victories that hastened the depletion of the fish stocks on which they depended.

Well-intentioned values that enrich our lives, transfigured by heroic myth and historical myopia, were pulled into regulatory debates over the fisheries with disastrous, albeit unintended, consequences. Over the past century and a half, Americans wrestling with how to manage New England's marine resources looked to archaic images of fishermen and fishing—not to the industrial, financial, or ecological realities—to inform decisions that ultimately affected the fishermen and the marine ecosystem. They managed the myth while the reality fell apart. This book addresses how New England's endemic objectification of fishermen and ardent belief in that timeless iconography shaped the meaningful regulation of a modern, dynamic, and innovative industry.

Reconstructing New England's heavy trawler fishery brings together three bodies of seemingly unrelated evidence. First, it relies on published federal fisheries data from the 1860s to the 1960s. Today, many fisheries scientists criticize such data for being inadequate by contemporary standards. Such critiques overlook the tremendous expertise, diligence, and consistency past data collectors brought to their work at a time when there were very few established standards. Officers in the U.S. Bureau of Fisheries and its future manifestations maintained and expanded data throughout the twentieth century, often extensively annotating it to explain changing data collection protocols and assumptions. While not as systematic or statistically rigorous as contemporary information, federal fisheries data for New England provides a rich, detailed archive revealing how government officials viewed the fisheries. To that end, I provide the weight of catch in pounds, as was usually reported before 1960 (and is familiar to most readers), and in metric tons, as is currently used in management (1 mt = 2,204 lbs.). Providing both seems a rational way forward.

I also use the terms "large," "medium," and "small" otter trawlers and draggers consistent with the data collected by federal port agents throughout the period. Given the differing sizes and industrial roles, I have used "heavy trawler" and "large trawler" interchangeably. The technical specifics differentiating vessel classes appear in the text, but in general, heavy trawlers were too large to be built by a group of local fishermen. On the other hand, a small or medium otter trawler—or dragger—could be built by an individual who had saved up enough funds. Heavy trawlers tended to be corporate-owned, while corporations as well as owner-operators owned draggers. Most importantly, heavy trawlers had a wider operational range and could fish for upward of a week or two. Smaller vessels stayed closer to home, for better or worse, as the century would reveal.

In addition to published fisheries data, this study also systematically explores popular representations of New England fishing during the same period to gauge how people perceived one of New England's most celebrated industries. These cultural sources come from popular novels, such as *Captains Courageous;* images like those of Winslow Homer; and, most importantly, newspaper and magazine coverage. More so than today, newspapers and magazines broadly shaped regional attitudes on many fronts, from politics to national identity. Commercial fishing, a regular feature in the region's largest newspaper, the *Boston Globe*, had an avid following. Yet coverage rarely changed as frequently as the industry changed, and newspapers sometimes helped New England residents misunderstand an industry that, while valued highly, most seldom saw.

The myths, realities, values, and management strategies evident in the scientific data and public discourse converged in an unexpected place: congressional committee hearings form the third body of evidence on which this work relies. In the United States, marine waters are a public commons that, once seized from Native Americans, were inherited legalistically—through war, revolution, and remonstrance—from the British Crown. As such, their management has, to varying degrees, always played out in a public forum. In the nineteenth and twentieth centuries, litigation and international negotiations over marine resource use necessitated keeping close records of those proceedings. In the twentieth century, transcripts of hearings, public meetings, and legislative debates swelled the documentary record.

Far from mere political theater, committee hearings represent, even today, a complex arena where committee members and remonstrants entwine the material, the ideological, the cultural, and the political in deciding the fate of public resources. Convened regularly, the meetings follow a predetermined schedule so that, if guided by an adroit chair, the work can be completed within the assigned timeframe.

Committee meetings are more efficient because they usually bring the same people together. This creates opportunities to present, test, evaluate, and revise rationales for granting or denying requests from the public. For witnesses, hearings allow them to present their positions, their evidence, and the reasons why their requests should be granted. For the politicians, hearings allow them to place into the public record why they support or oppose the proposals appearing before the committee. For sitting politicians serving local constituencies, making statements on the public record is not mere window dressing but the basis of future litigation. These statements represent the final calculus of the material, ideological, political, and accountable. As such, the reams of committee hearing transcripts reveal to historians how testifiers and committee members justified their choices to the wider public—in their own words.

I have assigned agency to ports and their representatives, not just to individuals. Gloucester and Boston were extremely active in fisheries politics throughout the period, whereas Portland, Maine, and New Bedford, Massachusetts, while significant fishing ports, rarely engaged in the regulatory process in the years covered here—at least not at the federal level. Furthermore, the two latter ports did not host many large trawlers, which formed the backbone of the twentieth-century New England groundfish fleet. From 1870 to 1960, fishing interests in Boston and Gloucester presented themselves as representatives of their homeports, places seen by many as units of change and agency. Federal officers collected data based on these assumptions. That said, the ports were adversaries as often as they were united.

What I hope emerges from the following pages is a history of a fleet and an industrial infrastructure quickly built on abundant stocks of haddock and redfish that disappeared just as quickly as the region and its political representatives conflated a modern, sophisticated, and politically engaged industry with a quaint, artisanal emblem of regional past glories. This story is important for no other reason than it provides a glimpse of how New England's fisheries evolved and devolved over the past century and a half. Although many might disagree, I believe it also helps us see that this pattern is unfolding again. If declining fish stocks in the mid-twentieth century sank the region's heavy trawler fleet, further declines are sinking the draggers that have managed to survive to the present. Maybe by exploring our region's fisheries' past through empirically driven historical analysis we can avoid letting our love for heritage, tradition, and myth kill off those fishermen who have managed to hang on.

My council experience suggests that if we wish to manage natural resources effectively, we must understand and acknowledge the empirically rooted history of how people have used those resources. Myth, imagery, and lore, standing in absentia for a thorough historical analysis of New England's industrial fisheries, have done as much to undermine fisheries resources as weak regulation, uncertain science, and profit-driven fishing interests. The difference is this: any historical examination must consider the cultural context within which fishing, or any activity, takes place. This means we—as a people—must examine our role in fomenting and perpetuating mismanagement. Refusing to see New England's fisheries for what they had and have become amounts to the public's abdication of our own stewardship responsibilities. In light of the cultural myopia enshrouding fisheries, we can no longer ask, "how did we get here?" Instead, we must ask, "why did we think we'd be anywhere else?"

CHAPTER 1

MOBILIZING A TRADITION

Gloucester and the Demise of the Schooner Fishery, 1866–1892

I n 1886, Gloucester's celebrated cod fishing industry saw a chance to right an outstanding wrong. Back in 1872, Gloucester's mackerel fishermen had successfully lobbied the U.S. government for a treaty granting the Americans access to Canada's inshore fishing grounds in return for the duty-free importation of Canadian fish products. The agreement, the 1872 Treat of Washington, pitted two of America's most important fisheries against each other. It was a good deal for the mackerel catchers, who were eager to exploit the rich coastal waters within three nautical miles of Atlantic Canada and Newfoundland. It was a bad deal for the venerable and much larger cod fishery, which faced new competition from imported Canadian salt cod in U.S. markets. Furthermore, the treaty permitted British naval inspection of American fishing vessels, and on the water, this meant that Gloucester's cod fishing schooners faced intensified search and potential seizure as they entered Canadian ports for supplies like food, water, or bait. Facing forfeiture on the high seas, market pressure at home, and marginalization at the hands of American mackerel fishermen, the owners of Gloucester's cod fishing fleet determined to redress these grievances when the treaty came up for renegotiation fourteen years later.

During those initial 1872 negotiations, the U.S. government itself had posed one of the biggest challenges to Gloucester's cod fishing outfitters—firms that owned and operated numerous schooners and fronted credit to the fishermen who worked aboard them. State Department officials believed that cheaper Canadian fish imports would benefit American consumers by providing protein to working families at lower cost. Gloucester's outfitters felt that this perception undercut the position of American negotiators and helped their British counterparts. They suspected that U.S. officials knew full well that the treaty would be detrimental to American outfitters but would agree to it anyway.

To lobby U.S. diplomats to support their position on the treaty, Gloucester cod fishermen and outfitters sent formal representatives, organized as the Gloucester Board of Trade, to Washington, D.C., just before the treaty was finalized. The meeting with State Department officials proved to be embarrassing and fruitless, however. According to Sylvanus Smith (who, although a member of the delegation, did not make the trip to Washington), Secretary of State Hamilton Fish refused even to see Gloucester's representatives, let alone hear their concerns. "He did not want to hear any fish stories," Smith recalled in his 1915 memoir. Instead, the delegation met with a junior senator who received them "very coolly" to accept any papers they might have with them. In the end, the Gloucester Board of Trade returned with nothing. Smith lamented that "this was the only time that a committee representing the fisheries has ever met with an absolute failure to present their case [to government representatives]."[1] As a result of British diplomatic adroitness and American bureaucratic indifference, Gloucester's cod fishing interests received little from the 1872 Treaty of Washington, a failure that continued to sting throughout the 1880s.

Gloucester's outfitters were determined that they would not be so easily dismissed when the treaty approached renegotiation in 1887. Organizing as the American Fishery Union, Gloucester cod fishing firms launched a formal political campaign in 1885 to prevent renewal. They invested time, money, and hard-won political savvy to make clear to State Department officials that American cod fishermen were unwilling to pay for the American mackerel fleet's access to Canada's inshore waters.

One critical element of their campaign was direct communication not with the State Department but with President Grover Cleveland himself. Referring to Canadian search and seizure of American vessels in international waters, the American Fishery Union reminded the president of his duty, "as the custodian of the honor and dignity of the flag in the foreign relations of this free and sovereign, and independent people of the United States, to protect that flag from the humiliation of being arrested in the high seas in time of peace, by foreign officers." The union implored the president to deploy U.S. naval forces to protect American fishing fleets working off Canada and demanded that Cleveland take a firmer stance in negotiations. In dealing with the professional British diplomatic corps, the organization cautioned President Cleveland "to put you on your guard," as the general opinion in Gloucester was that American representatives had been overawed by British diplomats in 1872.[2] Overall, the American Fishery Union's 1885 curt, condescending memorial demanded that the interests of Gloucester not be sacrificed for the sake of cheap imported food for American workers. The memorial emphasized that Gloucester's outfitters were watching and would engage directly should they believe their interests were on the chopping block again.

In addition to lobbying, perhaps even bullying, elected officials, the American Fishery Union pressed their case on a new front opening in the esoteric realm of international fisheries negotiations. Taking advantage of contemporary American discussions about national direction, the union appealed directly to the American public by using the press to build popular opposition to renewing the agreement. A former fisherman, federal census agent, and one of the industry's greatest supporters, J. W. Collins articulated the cod fishery's case before the American public. Writing in one of the nation's most popular magazines, *Century Illustrated Monthly*, in October 1886, he updated readers on the state of New England's fisheries. Collins's article, modestly titled "The Outlook of the Fisheries," details how commercial fisheries in the region employed over 130,000 workers, produced over $43 million in products, supported a population of more than half a million, and represented over $37 million in invested capital. Based on these figure alone, Collins wanted every

American to understand that the industry was "certainly entitled to consideration as an important factor in our national growth and prosperity." Importantly, Collins declared that New England's fisheries were more than just an industry. He took pains to inform his readers that its fishermen also constituted "a large body of hardy and enterprising men, who constitute a self-supporting militia of the sea, a force of inestimable value to any nation that aspires to naval or commercial greatness."[3]

Federal indifference, Collins believed, had sacrificed these patriotic New Englanders and their communities during the Treaty of Washington's original negotiation. Returning to ports he had visited while with the U.S. Census, Collins found "nothing to indicate [the fishing port's] former business importance but neglected and tumbled-down storehouses, and decaying wharves, against which lay a superannuated fish freighter, the tide flowing in and out of her open seams, and the broken cordage flapping monotonously against her bare spars, as if she had come here to die on the scene of her former usefulness."[4]

For Collins, previous U.S.-Canadian treaties had jeopardized an important element of American national character. As a result, the most recent round of negotiations had more at stake than economics. Indeed, the fate of many communities rested on the outcome of the talks. Those endangered communities encapsulated for Collins the best of America's national values: "Let our fishermen be once assured of protection in the enjoyments of their rights. . . . There can be no reasonable doubt that, with the improved methods and appliances [for fishing] which have been recently adopted, together with the bravery and hardihood which have been their distinguishing characteristics, the industry . . . will regain its former prosperity."[5] Fishermen did more than feed American bodies; they embodied the noblest of American souls.

Despite endorsements by President Cleveland, Canada, and Great Britain, the treaty was defeated in Congress, and forces far greater than Gloucester's outfitters decided its fate. Certainly, American Fishery Union activities helped incite New England's congressional delegation to kill the deal. Partisan politics and national debates over free trade and protectionism, however, also proved influential. Regardless of the means, the union ultimately got what it wanted: American markets closed once more to the free import of Canadian fish, and American producers free to charge the American people more for their products. Yet, with the defeat of the treaty, the union could claim that it had played a role in protecting local businesses, local fishermen, and an American way of life that many had come to see as a pillar of American culture—even though this victory constituted a loss for the public at large. Far from Collins's image of a struggling, humble body of simple, virtuous citizens, the union's actions revealed a group of enterprising operatives engaging politically and through the popular press to vocalize its positions on national affairs. Confident of its economic and political influence, at least in its own estimation, and cognizant of the costs of failure in 1872, the American Fishery Union willingly used the same techniques other American industrialists were employing to translate industrial strength into political, regulatory, and, in this case, diplomatic power. Like other major industries in the second half of the nineteenth century, Gloucester's outfitters leveraged industrial power to educate the president on how the obligations of his office could best be used to further Gloucester's international fishing interests.

More than obtuse industry lobbying in international tariff negotiations, the 1886 Treaty of Washington negotiations reveal an important, complex interweaving of politics, fisheries regulations, and popular culture in the late nineteenth century. Through Collins's writings and many other publications appearing about the same time, the general public began to see New England fishermen as something more than just workers on the water. His appeal defined the fishery, despite its expanding economic, industrial, and political significance, not as an industrial operation but rather as a social, cultural, and heritage-laden activity employing some of the American republic's best citizens. While readers of the *Century Illustrated Monthly* living outside of New England likely cared little for New England fishing, middle-class Americans had begun to see industrialization, urbanization, and immigration as threats to American values by the 1880s.[6] As an integral part of an idealized past, mariners and fishermen in particular came to carry greater cultural significance.[7] As the American public grew more concerned about the fate of these human vessels of national culture, they came to see fishermen as one of the last vestiges of a perceived innocent and harmonious antebellum past.

Although arcane and somewhat wonkish, the 1872 and 1886 treaty negotiations also reveal how Gloucester's outfitters quickly learned to engage in political processes in sophisticated ways. Their willingness to do so encapsulates how industrial sophistication and cultural representation of the postbellum New England fishery did not stand in isolation. From the 1880s to the 1960s, New England's largest fishing interests pursued market power and profitability through industrialization, trusts, combinations, cartels, mechanization, union-breaking, consolidation, lobbying, public relations campaigns, and corruption. By 1925, one of the most powerful fisheries to emerge from those processes, Boston's fleet of heavy haddock trawlers, resembled other major American industries that sought to optimize fiscal performance using all means, good or bad, that characterized industrialization during the period. Yet unlike the steel, railroad, oil, auto, and other major industries, New England fishing interests enjoyed a unique and powerful asset—their idealized image. Americans consistently viewed this industry as an emblem of national "tradition," "heritage," and "timelessness," whereas in reality it was a modern, sophisticated, and dynamic combination of money, markets, technology, workers, and marine resources. New England's offshore fisheries used this idealized image to powerful advantage as they pursued greater profitability over the following century.

Competition, Consolidation, and Industrialization in the Gloucester Fleet

New England's antebellum cod fishery developed within a protected market and benefited from one of the first industrial subsidies in U.S. history. The 1789 Salt Bounty, passed by Congress shortly after independence, paid cod fishermen a rebate, based on vessel size instead of catch, to offset import duties on the salt needed to cure their harvest. From its inception through the Civil War to 1866, this direct subsidy evolved into an important element that shaped New England fisheries. The federal government paid the bounty in cash, whereas sale of the catch to merchants was usually on credit. Moreover, the bounty was paid directly to the fishermen, often in lump sums at the end of the season. Thus, the bounty's liquidity and timing spread necessary cash throughout New England's shore communities in early winter, when it was most needed. While it is difficult to measure the impact of the Salt Bounty on the development of antebellum New England's coastal economy, it was likely significant, if not essential, in more remote, cash-strapped fishing communities along the Gulf of Maine.

The bounty's influence on the fishing industry extended beyond well-timed cash payments. Qualifying criteria determined who fished aboard New England vessels and how they were paid, and which fish were landed and how they were sold. Only vessels carrying a majority of American fishermen paid on shares became eligible for bounty payments. The share system, in place since the seventeenth century, treated each fishing voyage, or fare, as an independent commercial partnership among all fishermen working the vessel and the vessel owners. After overhead costs for food, ice, dockage, offloading, catch processing, and bait was reduced from each fare's total revenue, the vessels owners received a predetermined cut, and from the remainder, each fishermen received pay based on their individual contribution to the total catch. While the system allowed fishermen to enjoy the proceeds of windfall fares, it also privileged the vessel owner's payment as overhead, even when a fare failed to clear enough to pay fishermen for their work. As a result of the bounty's direct payments to fishermen, while other industries embraced more modern forms of labor compensation, the antiquated share system remained as the principal wage structure in the New England fisheries (as it remains today).

Furthermore, bounty requirements mandated that vessels could only fish for cod that was to be salted and dried for sale. Operationally, fishermen caught and cured cod during the spring, summer, and fall, leaving them time for other work during the rest of the year. Selling fresh cod violated the law, as did engaging in another fishery during the stipulated 120 days of fishing. These provisions stifled the development of other fisheries and kept the majority of investment in the salt cod fishery, even as urban immigration increased the demand for other fish products before the Civil War. Bounty requirements also hindered the flow of immigrant labor into fisheries with its requirement that three-quarters of every fishing crew had to be American citizens. Finally, the bounty placed a premium on literacy, as captains were required to keep logbooks and submit their vessels to annual inspection.[8]

For many, these requirements were well worth the hassle. With five-eighths of the bounty payment divided among the crew, and three-eighths going to the owners, federal cash payments provided enough revenue to offset operational costs for vessel owners and to attract and retain American crews far from the fishing centers of Boston, Gloucester, and Salem/Beverly, Massachusetts, and Portland, Maine.[9] As a result, New England's antebellum fisheries came to include major fishing centers, like Boston and Salem/Beverly, as well as numerous small-scale operations scattered along every nook and cranny of the New England shore. Reliant on the bounty, local capital, local labor, and local markets, New England's antebellum fishing operations followed a seasonal routine in fishing entrepôts and outports alike.[10]

After the Civil War, however, opportunities offered by postbellum industrialization in Boston and Gloucester provided new avenues of growth independent from government subsidies. The war's intensified business climate played a critical role, as industry leaders in these major fishing centers began to see the market as a zero-sum game in which competition had to be eliminated. Gloucester's leading fisheries outfitters had no problem using political strategies to winnow out competitors. In 1866, they concluded that the Salt Bounty's restrictions had hindered their own operations and propped up outport cod fisheries long enough. When congressional representatives from southern states challenged the bounty in 1866, as they had done for decades, Gloucester's leaders requested—in a departure from the past—that their congressional delegation allow the law to die.[11] Deprived of the federal cash subsidy, small owner-operated fishing operations throughout the region began to suffer as Gloucester's outfitters engaged in a frenzied attempt to dominate New England's fishing business.

Nor were Gloucester's outfitters willing to stop with the repeal of the bounty. From the end of the Civil War to the end of World War I, cod outfitters worked every angle to marginalize all remaining competition. Integrating operations vertically and horizontally helped them corner the American salt-fish market. As outport fishing operations declined, so poignantly described in Collins's 1886 piece, Gloucester's firms wrestled with one another to be the last one standing. This struggle emerges clearly from vessel ownership patterns detailed in the port's annual *List of Vessels Belonging to the District of Gloucester* between 1869 and 1908.[12] As early as 1869, only three years after the bounty's repeal, 404 of the 435 vessels over twenty tons in all fisheries that sailed out of Gloucester Harbor proper were owned by one of the port's fifty-three corporate outfitters and operated by hired captains and crew (about 93%).[13] By 1872, the percentage of corporate-owned vessels had fallen to roughly 70 percent, and stabilized between 60 and 70 percent until 1908. Although the proportion of vessels owned by outfitting firms declined during this period, the number of firms declined more rapidly due to the intense competition. From a high of fifty-three in 1869, four years later only thirty-eight outfitters remained. A rise in the mid-1880s, perhaps reflecting optimism over the end of reciprocity with Canada, proved short-lived. By 1908, only twenty-one firms still ran schooners. Seven years later only two of the fifty-three businesses operating in the 1860s remained.[14] From the end of the Civil War until the end of World War I, the Gloucester fleet had experienced industrial consolidation as intense as any contemporary American industry.

However, despite the repeal of the bounty and consolidation in the fisheries, small-scale operators never disappeared entirely. In 1887, for example, A. Howard Clarke of the U.S. Nation Museum (Smithsonian) in Washington reported—perhaps optimistically—that "it is gratifying to note that the number of vessels owned by fishermen themselves is rapidly increasing."[15] Yet, momentary upswings in independent, fishermen-owned and -operated vessels rarely lasted. As smaller operators came and went, and smaller outfitters rose and fell, large firms persistently filled the voids, a trend revealed by comparing the average and median numbers of vessels owned by each firm. After a high of about nine in the anomalous year of 1874, the average number of vessels each firm owned fluctuated around seven until 1908. However, the median number of vessels owned by each firm declined from seven in 1880 to five in 1908. While the averages remained relatively constant over time, the few firms that owned more vessels masked the increasing number of firms that owned fewer vessels.[16] These statistics illustrate the marked consolidation in Gloucester's fishing operations between the 1870s and 1890s as large firms seeking to dominate the fisheries pushed out smaller operations. And by the first decade of the twentieth century they had done just that. In 1908, so few outfitters remained that one firm, John Pew & Sons, secured the capital to buy out its last salt fishery competitors once and for all. Allied with investors from outside the fishery, Pew & Sons formed the Gorton-Pew Fisheries Company. Within a year, the firm tripled the size of its fleet from fifteen to forty-six vessels. Like other highly capitalized American industries, Gloucester outfitters responded to market dynamics not by embracing competition but by curtailing it.

Few native-born Americans were willing to work in an industry marked by such an intense business climate and ruthless competition. In the decades following the Civil War, the majority of fishermen working in Gloucester's fisheries were foreign-born. Between 1869 and 1908, the annual *List of Vessels* identified birthplaces for 946 of the 1,074 fishermen who drowned when vessels went down with all hands. If those tragedies occurred at random, therefore providing a grim snapshot of who actually worked aboard Gloucester's fishing vessels during this period, then the fleet was predominantly Canadian. Of those lost fishermen, 577 (61%) were born in Canada and Newfoundland. In stark contrast to the unique American character ascribed to New England's fishermen, only 118 (12.5%) were born in the United States. The remainder came from various European countries, with a few from the West Indies. Of the 118 American-born fishermen, 31 (3.3%) came from Gloucester, another 40 (4.2%) were born in the rest of Massachusetts, 38 came down from Maine (4.0%), and 9 (1.0%) were born elsewhere in the United States. Hiring practices drove these figures. Gloucester captains often stopped at Canadian ports for additional crew on the voyage out and dropped them off on the return. While many of the Canadians working on Gloucester vessels never touched American soil, the fact remains that only a handful of Gloucester's native sons went to sea when the town's most famous industry flourished.[17] As soon as they were freed from the bounty's restrictions, Gloucester's outfitters, like other large American industries, took full advantage of foreign-born labor.

To contemporary observers the trend was clear. At a time of virulent nativism, numerous government investigations, including those conducted by the Census Bureau, the U.S. Fish Commission, and the Massachusetts Commissioners of Fish and Game, all commented scornfully on how many Nova Scotians and Newfoundlanders worked on Gloucester vessels. The large Canadian presence in American fisheries also caught the attention of newspaper reporters covering the U.S.-Canadian fisheries disputes.[18] Ironically, while deriding the jobs lost to American workers, or the American money flowing out to foreign, largely Canadian communities, official reports and newspaper stories actually underestimated how many Gloucester fishermen hailed from foreign ports. Morbid statistics of lost fishermen indicate that Gloucester's schooner fishery was as much a colony of Canadian workers as a pillar of American capitalist enterprise.

As outfitters consolidated their power, fishermen attempted to do the same thing. Late in 1885, as vessel owners developed political strategies to defeat the Treaty of Washington, crewmembers formed the Deep Sea Assembly of the Knights of Labor explicitly to counter the power that outfitters and vessel captains wielded over the production process. The *Boston Globe* reported that "the fishermen of Gloucester are impressed with the idea that monopolies and rings exist in the fish business . . . and there is a labor element strong enough to accomplish reforms in the fisheries, if those interested band together under the Knights of Labor."[19] Their effort proved remarkably successful. By March 1886, the union had enlisted between one hundred and two hundred members. It had also grown some teeth. The Deep Sea Assembly challenged assertions that only outfitters and their political allies spoke for Gloucester's fisheries. Although Gloucester's "fish trust or monopoly" opposed renegotiating the treaty with Canada, the Knights endorsed a series of resolutions supporting the president's diplomacy and criticizing Massachusetts senator Frisbie Hoar for his ignorance of fisheries business. In particular, the union objected to Hoar's repetition of claims by major outfitters, such as Sylvester Cunningham, that fishermen universally embraced the shares system of crew payment. "Senator Hoar, while talking for American fishermen, really represents the vessel owners," union members claimed. "Not only the Knights, but fishermen generally insist that Senator Hoar does not understand the subject. . . . All the information possessed by Hoar had come from the owners."[20]

When it became clear that the Salt Bounty kept large firms from adapting to changing market conditions, outfitters let it lapse. Nevertheless, the passing of the bounty unleashed industrializing forces largely alien to fisheries before the Civil War. While outfitters enjoyed some of the same opportunities that late nineteenth-century American capitalism offered other industrialists—plentiful foreign labor, looser regulations, and unhindered competition—they also entered a frenetic ring of business-on-business pugilism. In the fray, fishermen formed unions to protect themselves. There was little protection, however, for the smaller operations scattered along the New England coast that once relied on the bounty for solvency. During the tumultuous fifty years from the end of the Civil War to the beginning of World War I, outfitting firms that could harness the forces of industrialization cannibalized firms in fishing centers or outports that could not do the same. Although still powered largely by sails and oars, and fishing with hooks and lines, at the turn of the twentieth century, Gloucester's fishing fleet had become a modern, highly capitalized industry defined by competition, consolidation, foreign-born labor, and union organization.

Constructing New England's Fishermen

Few outside the industry noticed this profound transformation. Between 1885 and 1912, a fundamental disconnect emerged between popular perceptions of New England's fisheries and its industrial realities. Ironically, outside observers painted with the same brush the small-scale, outport family fishing operations and the large-scale, industrialized Gloucester firms whose sophisticated acumen threatened their smaller-scale rivals. By the 1890s, the bucolic imagery of New England outport fishing had so thoroughly displaced the industrial realities experienced by Gloucester outfitters that it added a blush of virtue to their legislative and regulatory agenda, albeit initially in a confused and contradictory manner. Not until 1912, after full-fledged mechanization, would contradictions between the cultural representations of fisheries and their industrial reality come into focus.

J. W. Collins, Gloucester's public relations ally in the 1886 campaign, was not the first person to celebrate the historical and social value of New England fisheries. Many commentators focused on the industrial and economic evolution of the fisheries and offered cultural accolades only in passing. Thomas Jefferson, Humphrey Storer, Lorenzo Sabine, Spencer Baird, and others, in government reports and disquisitions after 1790, tied New England's fisheries to Europe's first North American settlements and the founding of the American republic. In 1882, for example, government researcher and census analyst W. A. Wilcox argued in *Forest and Stream* magazine that, while fishing was one of New England's oldest industries, modern operations had adopted new catching technologies, expanded shoreside processing plants, enhanced preservation techniques, and extended distribution systems to maintain and expand markets. Yet far from moribund, as Collins would have it, Wilcox concluded that, with the fleet concentrated in fewer ports, New England fishing was "as productive at the present time, with half the number of vessels engaged, as in past years."[21]

By 1886, when Collins took pen to paper, however, New England fishermen had become an unlikely American *cause célèbre*. Two events thrust them into the public eye at a pivotal moment in American cultural discourse. In the first, the heroic exploits of Gloucester halibut fisherman Howard Blackburn, publicized in 1885, highlighted the virtues ascribed to fishermen that many saw waning in industrializing America. The details of Blackburn's experience make it one of American maritime history's most remarkable incidents. In January 1883, Blackburn brought his dead dory-mate's body ashore in Newfoundland with his hands literally frozen to the oars. The two had been fishing from a dory off the schooner *Grace L. Fears* when a winter storm separated them from their vessel. Over the next five days, Blackburn rowed the boat toward shore as his dory-mate slowly succumbed to exposure. Determined to give his dory-mate a proper burial on land, Blackburn endured winter storms, freezing salt spray, and cruel exposure before reaching safety in Newfoundland. When Blackburn returned to Gloucester, the city gave him a hero's welcome, and the ensuing private donations allowed him to establish his own fishing business. Not surprisingly, this heroic story of endurance and dedication made headlines around the country: newspapers as far away as Kansas City (where it made the front page) and St. Paul, Minnesota, ran the story in January 1885. J. W. Collins entered the ring again when he wrote up and published Blackburn's complete account and marketed it up and down the eastern seaboard.[22]

At the same time, Americans found themselves reconsidering New England fishermen visually. In the second event, Winslow Homer unveiled four paintings featuring New England fishermen at a prominent art show in New York in 1885. Not only did the paintings demonstrate a technique that impressed Homer's fellow artists; his choice of subject and heroic portrayal of common American workers won him wide popular acclaim. The most celebrated painting of the series was related to Blackburn's account. *The Fog Warning* portrays a lone halibut fishermen pulling through tempestuous seas for a distant schooner sailing away.[23] It provided visual confirmation of the hardiness of New England's venerable industry. Instead of portraying fishermen struggling in vain against the fathomless sea, the convention in earlier schools of marine painting, Homer depicted a single fisherman, strong, resolute, and brave, humbled but not bowed by the natural forces surrounding him.[24] Like the Civil War etchings that first drew attention to his work, Homer put a common man's face onto a larger American story, appealing to the viewer on personal and emotional grounds as well as on cultural values and artistic merit.[25]

Blackburn's heroism and Homer's acclaim came at a time when many Americans wondered where their country was heading. By the 1880s, middle-class Americans had begun to see threats to American values along this gilded path.[26] Many Americans, especially white Americans in the Northeast, wondered how their vision of the United States would survive the social and cultural pressures wrought by rapid industrialization, immigration, urbanization, and political consolidation. For many, New England fishermen, or more accurately popular representations of them, often symbolized what had been good and what might be lost if America continued down the industrial path it had chosen. In an integral part of the discourse on national values, mariners and fishermen in particular came to be seen as links to, and defenders of, a colonial past ascribed with a distinct set of conservative values.[27] Historian Glenn M. Grasso defines this cultural moment as the "Maritime Revival," which grew out of the contemporary Colonial Revival aesthetic. He argues that "Old-stock Americans responded to [industrialization] by locating core values in pieces of history such as the Revolutionary generation, early colonists, architecture and seafaring."[28] Similar trends in Atlantic Canada, analyzed by Ian McKay, identified national character as originating from mariners and, in particular, from Atlantic Canada's fishermen.[29] As the American public grew more concerned about these human vessels of national culture, fishermen came to be seen as vestiges of that innocent and idealized past. As historian Bruce Robertson argues, by the late nineteenth century Americans had begun to see New England fishermen as "sentinels, guardians, the first immigrants to the New World, and the last chance against change."[30] Indeed, representations of fishing that depicted the work as unmodern, uncapitalistic, timeless, and unchanging acquired greater cultural valence as American underwent industrial change.

> What gave these images their special strength at the end of the nineteenth century, and their special location in New England, was the particular collision between history and modernity enacted there and then. New England, by the end of the century, claimed more convincingly than any other region to be the place where the acts and ideas that had engendered the United States had come into being and where living remnants of that authentic history still survived. As the notion of what constituted the heart of New England receded from its southern tier of states, Boston and points north promoted their claim to stand apart from the ongoing movements of American history and rebuke its excesses and probable future. Against the endless and ambiguous future, New England was presumed to represent the constancy of timeless values, a strong rock amid a sea of change. History might have its domestic, genteel, and variable side, but it could also embody virile political ideals—Revolutionary values of independence, action, strength. This version of history found its living expression most popularly in images of seaboard communities. The New England coast was granite of a particularly principled kind.[31]

The contemporary popular interest accorded to Blackburn and Homer certainly supports this conclusion. Through their experiences and images, and those of other New England fishermen, Americans interpreted the changes affecting the region's fisheries. Few people cared about fish tariffs. When personified by Blackburn and Homer, however, and presented in human terms, as J. W. Collins had presented the tariff debates, fisheries policy could engender strong feelings even among casual observers, compelling them to care about the fishermen, their fate, and the fate of the nation. By conflating modern fears of social change with romanticized images of an idealized maritime past, the producers of these images unintentionally created a powerful political tool, used and eschewed as time and circumstance dictated by fisheries advocates and opponents alike. After 1885, fishing was no longer just another American industry: to many conservatives it had become the benchmark of American continuity.

Politicizing, Appropriating, and Refining a Cultural Icon

As writers and painters wrapped Gloucester's industrialized fleet in the mantle of tradition, New England's outport fishermen did not sit idly by. In southern New England, state and local campaigns fought against weirs and pound nets—capital-intensive gear that was marginalizing day-boat fishermen. Despite these challenges, outport and day-boat fishermen across all of New England in the 1870s and 1880s remained largely silent on the fisheries debates in Washington, D.C. Parochialism came at a price. By the 1890s, the "traditional" outport fishing operations that most often illustrated New England's fishing tradition found themselves sidelined by well-capitalized competitors in regional fishing centers.

Debates on the 1891 Lapham bill finally mobilized New England's outport fishing communities to fight against a monopoly's bald-faced gambit to usurp states' rights to regulate their fisheries in inshore waters. The bill essentially sought to remove from the states and grant to the federal government the authority to regulate state waters (that is, waters up to three miles from shore). Supported by large-scale, politically connected fishing interests, the bill would move the center of decision-making to a venue where those interests held great sway over lawmakers. One organization in particular pushed for the bill. The U.S. Menhaden Oil and Guano Association (USMOGA), representing large menhaden rendering firms along the eastern seaboard, wanted to see the bill passed so they could more easily change current regulations preventing them from fishing within state waters.[32] In response, outport fishermen in southern New England pulled together a coalition of conservation-minded lawyers, local fish dealers, and the Maine State Commissioner of Fisheries to fight the conglomerate's political machinations. In the ensuing struggle, small-scale fishermen used the powerful symbolic imagery of New England's family fisheries to combat the threat from the large fishing corporation. In doing so, they introduced a new tool to domestic fisheries politics. While USMOGA presented economic arguments as to why states should abjure inshore regulation, allied outport fishermen from southern New England to the Maine coast invoked social and cultural images highlighting the social aspects of fisheries management. Against teams of lawyers and lobbyists, representatives from Maine's and Massachusetts' inshore fishing operations challenged elected federal officials to consider the consequences of their decisions for the very fishermen many Americans believed to personify imperiled American virility and cultural values.

The struggle against the Lapham bill once again stemmed from consolidation within the fishery. Forty years earlier, the New England menhaden fishery had been a decentralized affair. Small-scale local operators fished menhaden for bait, fertilizer, or oil. During the Civil War, however, operations in southern New England consolidated as the more successful firms invested in larger fish processing facilities. After the war, these firms moved into the Mid-Atlantic states, taking advantage of cheap African American labor hungry for work in the Reconstruction South.[33] Finding quick success, the menhaden fishery embraced mechanical propulsion by building the first expensive steamers in the American fishing industry in the early 1870s.

As with Gloucester's cod fishery, the menhaden fishery consolidated, capitalized, and grew more sophisticated. The remaining menhaden operations, large processing factories, and steam-driven purse-seining vessels (which encircle whole schools with large nets), representing unprecedented investment in a fishery, formed the USMOGA cartel in 1889 to manipulate prices for oil, fish scrap, bait, and fertilizer.[34] Shortly after Gloucester's American Fishery Union lobbied Congress to stop imports of cheap Canadian salt cod, USMOGA used its economic clout to lobby against state fisheries regulation. A series of laws passed from Massachusetts to New Jersey in the late 1880s banned the use of purse seines within state waters and severely restricted the menhaden fishery. In response, USMOGA members devised a strategy to nullify these regulations. A menhaden operation in Tiverton, Rhode Island, the Daniel T. Church Company, sent Captain Arthur Manchester out to intentionally break the new laws. With this test case, USMOGA could challenge states' rights to regulate inshore waters in federal courts. Arguing that creation of the U.S. Fish Commission in 1871 had stripped states of their rights to regulate waters within three miles of shore, Manchester relied on USMOGA attorneys to successfully appeal his conviction all the way to the U.S. Supreme Court.

In Washington, however, the case did not fare well. The Supreme Court quickly upheld Manchester's conviction and rejected USMOGA's arguments in *Manchester v. Massachusetts* on almost every level.[35] The cartel would not be deterred, however. USMOGA launched a legislative campaign to succeed where its legal campaign had failed. In 1892, the conglomerate enlisted Rhode Island congressman Oscar Lapham to sponsor a bill that would strip states of their right to regulate fisheries inside three nautical miles of the coast. In the bill, regulating fisheries would devolve to the U.S. Fish Commission in Washington, where it would come under the influence of national politics and Gilded Age lobbying. In effect, this stripped outport fishermen of their voice in shaping local fishing laws.

From the start, USMOGA used every political tool at its disposal to push the Lapham bill through. According to Massachusetts attorney Charles Francis Chamberlayne, an ardent opponent, as the bill went into hearings before the House of Representatives' Committee on Merchant Marine and Fisheries, USMOGA paid newspaper hacks to pen supporting pieces and bribed editors to time publication so as to maximize their influence. USMOGA used the popular press more insidiously than Collins had done earlier. The cartel also resorted to blunt, thuggish tactics. Again according to Chamberlayne, they threatened a Massachusetts Fish and Game commissioner with personal and professional ruin if he continued to oppose the bill. Chamberlayne additionally claimed that the committee chairman was on the take, convening meetings once a week for an hour at a time so as to drag out the proceedings and increase costs for the bill's underfunded opponents. The association went so far as to pay off government stenographers so that the Congressional Record would be, as Chamberlayne characterized it, "materially altered and masculated [sic]."[36] Here was a consolidated fishing industry using all the economic and political power at its disposal to strip states of the right to regulate their inshore waters.

Corrupt or not, USMOGA's tactics were working. The only two people fighting the bill, Chamberlayne and New Bedford fish dealer George H. Palmer, saw defeat looming. The inordinate expense of living in Washington, D.C., week after week, waiting for a one-hour hearing, soon exhausted Palmer's funds, and he returned to New Bedford. Desperate for new allies, Chamberlayne recalled that the Maine Fish and Game commissioner, Dr. Edwin W. Gould, had "firmly rooted in his mind, that it was part of the duty of a Fish Commissioner to preserve fish as well as propagate them." Chamberlayne sent word, and Gould came to Washington to join the fight.[37]

When Gould arrived, "the battle seemed clearly a losing one," as Chamberlayne recalled to the American Fisheries Society in the aftermath of the affair. Arrayed against Chamberlayne were USMOGA lawyers, supported by lobbyists from fertilizer companies, fishing net manufacturers, oil industries, cotton and shoe trades, fish wholesalers' associations from Boston and across the Northeast, and the Gloucester Board of Trade. Also attending the hearings were representatives from the U.S. Fish Commission and, without clear association, J. W. Collins himself. Gould and Chamberlayne saw few friends in the gallery and found little support in public comments. The committee and the press received "industriously and discriminately employed" editorials, town resolutions, and private letters supporting the bill even before hearings began. Some even promised "political annihilation" if the bill failed.[38]

Against this array of powerful opponents, Gould and Chamberlayne decided to challenge the bill's constitutionality, a strategy supported by the court's decision in *Manchester v. Massachusetts*. By framing the bill as an attack on states' rights vis-à-vis federal authority, they would likely attract support from governors, state fisheries directors, and other constituents outside Washington concerned with the national aggregation of power at state expense. Their strategy also enabled rapid response. Lacking the time to mobilize widespread public opinion, it focused instead on civil servants and elected officials who could rapidly come to their aid. Soon after Gould's arrival, he and Chamberlayne mailed their constitutional case to governors and state fisheries commissioners across the United States and its territories, to congressional delegations in Washington, and to state officials as well.

Gould and Chamberlayne added to their argument over constitutionality emotional appeals highlighting the human consequences at stake, just as Collins had done in 1886. Written on behalf of the entire Maine Commission of Sea and Shore Fisheries, Gould's *Memorial Relating to the Destruction of State Fisheries* argued that the Lapham bill was as much about the duties of government in an industrial age as it was about industrial development. This pamphlet made two fundamental arguments in its critique of postbellum American industrial capitalism. First, Gould targeted the disproportionate power a few threatened to wield over the many: "The gentlemen who have appeared before the Honorable Committee [to support the bill] have represented, by comparatively few men, millions of capital wielded with the easy flexibility which the organization of the United States Menhaden Oil and Guano Association enables a few men to wield it." For Gould, the issue transcended dollars and cents and rested squarely on the question of for whom should government work. "The [Maine] Commission of Sea and Shore Fisheries feels itself charged with the conservation of interests, which, while largely exceeding in aggregate money value the assets of the Association, are [shared with] those people so numerous and often of such humble circumstances as to preclude their representation." The bill may have made economic sense, but it undermined Americans' understanding of the just rewards for hard work. Gould also invoked the cultural image of the New England fisherman as representing the reality of Maine fishing communities. He argued that the social and cultural consequences of the bill would undermine the labors of a humble, virtuous, even heroic fisher folk who would be dislodged from the seas:

> [The Commission of Sea and Shore Fisheries] would wish to present the cause, in the first place, of thousands of hardy boatmen and fishermen of our coast.... For these men, such legislation as that proposed by the "Lapham Bill" would strike down the strong arm of the State, which is raised, as it were, to defend them against a most merciless and crushing competition by combined capital, against whose improved methods and gigantic appliances the simple energies and rude devices of the average poor fisherman leave helpless and hopeless.[39]

Taken together, the growth of disproportionate power and the dislocation of American workers boded ill for the nation as a whole. Maine's outport fishermen were, for Gould, "hardy, fearless, inured to hardship and accustomed to brave the elements at all seasons, [and as such] are an ideal body of men." They embodied ideal citizenship, in that "they are essentially a generous class of men who do not wish any law enacted to discriminate in their favor, but who desire to have all citizens of the United States share with them, all being treated alike and subject to the same restrictions." Despite that egalitarianism, Gould believed that in a world where a select few enjoyed such power over so many, those embracing the admirable traits of hard work, honesty, and integrity were entitled to special consideration: "But for the care of these men, among her most valued citizens, our State has always shown extreme solicitude, and the force of facts which would crush so splendid a race of men should receive careful scrutiny.... The Nation has a manifest interest in the welfare of this class of men."[40]

Gould's *Memorial*, submitted to the committee in March 1892, redefined the terms of the debate entirely. No longer was the Lapham bill merely a constitutional matter out of which would emerge abstruse details concerning relationships between states and federal powers. While Chamberlayne argued the constitutional issues, Gould debated the costs the nation would incur should such changes be measured only in monetary terms. Gould and Chamberlayne turned the bill's deliberations into a debate on the future direction of America's political economy.

By framing the issues in broad, moral terms, the *Memorial* provided prospective allies with the political cover they needed to oppose the bill. Massachusetts governor William Russell dispatched the commonwealth's attorney general to support the opposition. More help came from Michigan, New York, Virginia, and Maryland, all of whom saw how the Lapham bill would hurt state interests. Ultimately, the combined forces of state officials, fisheries directors, and governors, under Gould's direction, helped defeat the bill in committee. With characteristic tenacity, however, USMOGA pushed another version "still in more dangerously insidious form," according to Chamberlayne's recollection, which passed in a six-to-five vote. In a feat of parliamentary maneuvering, however,

Gould arranged to kill the new version by compelling one of the "yes" voters to move for its reconsideration when its supporters were absent: the new version was killed by a unanimous six-to-zero vote.[41]

The defeat of the Lapham bill was important in a number of ways. The bill would have forced outport fishermen to oppose yet another powerful, monopolistic fishing industry seeking to squeeze out economic and political competition. This time, the results showed, outport fishermen and their allies were capable of adopting sophisticated lobbying tactics, including targeted political campaigns, to achieve legislative victories in Washington, D.C. Like Collins, Gould and Chamberlayne eloquently presented their case to a wider audience. Only the curtailed congressional hearings schedule prevented them from going to the popular press as Collins had done a decade earlier.

Most importantly, however, in reframing the Lapham bill, Gould further refined the idealized image of New England's fishermen. Noble, brave, hearty, honest, hardworking, and imperiled—these attributes would be associated, intuitively or intentionally, earnestly or tactically, to New England's fishermen and the debates about the fisheries' regulations for the next century. Outport fishermen reclaimed in the 1890s the imagery created by Collins and Gloucester outfitters in the 1880s to defeat USMOGA and its allies, including the Gloucester Board of Trade.[42]

Industrialization without Mechanization

During the last third of the nineteenth century, New England's fisheries, in particular Gloucester's cod fishery, underwent dramatic change. Like other heavy industries in America, the fisheries consolidated in a tumultuous and combative process. They took advantage of plentiful immigrant labor, market power, and economies of scale to absorb their rivals and survive—or perish. As the twentieth century dawned, they had acquired political clout through economic influence and learned to embrace more sophisticated approaches to regulatory and political influence.

New England's fisheries experienced another metamorphosis seemingly antithetical to their industrialization. In the course of thirty years, the fisheries went from being seen as a modernizing American industry in the Gilded Age to emerging as an emblem of America's heroic past and virtuous national character, a bastion of timeless values confronting modernization and industrial capitalism. In doing so, the nation imbued New England's fishing industries with the cultural traits and values many Americans feared were at risk. These two restructurings—one industrial and one representational—emerged in dialogue with one another. As fisheries embraced modern business practices, they used all available tools, including, ironically, their new iconic status, to affect changes that would facilitate consolidation and industrialization.

CHAPTER 2

THE BENEFITS OF MODERN FISHING

Boston and the Birth of Heavy Trawling, 1905–1925

If fisheries debates in the 1870s and 1880s had established a powerful romantic image of New England fishermen, technological change and partisan politics had spread it far and wide at the turn of the century. This time, however, Gloucester's cod fishing outfitters suffered striking defeats at the hands of a new, innovative, and efficient fishing industry emerging out of Boston's fish dealers. While Gloucester salt fish firms battled among themselves to own the schooner fishery on Cape Ann, Boston's fresh fish dealers expanded their control over all other New England fish markets. They used their considerable investment capital to recast New England fishing as a modern, mechanized industry. In doing so, they confronted and utterly defeated what remained of the American Fisheries Union and its allies. Nonetheless, Boston's dealers soon discovered that even a modern and mechanized industry needed to wrap itself in the mantle of tradition to operate free from scrutiny.

Dominating the Boston Fish Markets

Like Gloucester and the southern New England menhaden fishery, the story of Boston's fisheries features corporate consolidation and increasing industrial power. In Boston, spatial control of the waterfront, and the nagging fact that ice melts and fish rots, played critical roles in translating industry scale into political influence. New England's largest city, Boston, had long hosted fishing and supported one of the nation's largest markets for fish of all sorts. Indeed, in the colonial period it was an important northwest Atlantic hub for the transshipment of regional salt fish cargoes to markets across the world's oceans. Yet little more than a century later, the city pursued fishing very differently from Gloucester and the outports. Boston's large population lived close to the waterfront and increasingly liked to eat fresh fish. Not surprisingly, throughout the eighteenth and early nineteenth centuries, Boston fishermen sold fish fresh from their boats directly to consumers. In winter, when temperatures dropped below freezing, dealers also exported fresh fish as far as Albany and Montreal. Taste and freshness soon came to define Boston's fish market.[1]

At first, fresh-fish fishing could not be consolidated as easily as the salt fisheries. Local markets bought fish directly from small vessels working close to home that landed their catches daily. For these fishermen, selling to the consumer or retailer directly minimized the significant overhead costs the banks salt fisheries faced, in particular procuring and outfitting large vessels for distant water fishing and paying the shoreside middlemen processing fresh fish into salt fish and distributing the product to buyers. By 1835, however, as Boston's population grew with immigration and industrialization, increasing demand for fresh fish had created new business opportunities. The city's first fresh fish wholesale operation, Holbrook, Smith & Company, consolidated offloading fish from Boston's numerous day-boats to open the first dedicated wholesale fresh fish market in the city. Three years later, the firm expanded into a new facility on Commercial Wharf, a location that would become the center of the city's fresh fisheries for the next fifty years.[2]

As Boston's fresh fishery grew, power consolidated, not in the hands of vessel owners and outfitters, as had happened in Gloucester, but in the hands of the dealers and wholesalers who controlled how fast fresh fish moved across the dock from boat to buyer. With a catch's freshness fading as ice melted, whoever controlled the pier controlled the fishery, and that fact presented independent fishermen with a significant barrier: if they wanted the best price for fresh catch, they needed to unload the fish as soon as possible. Fishermen also faced another problem. Fish buyers interacted daily as they bid against each other at auction, and in their tightknit group, cutthroat competition made little sense. Soon buyers acknowledged their common interest in paying as little as possible for the fish and selling them for a lot more. Fishermen who were desperate to move their catch quickly provided wholesalers with an opportunity to leverage the most out of the market.

As a result, from the 1860s through the 1880s, dealers around the region used their control over markets and fish piers to enhance their profits. One way was for fish dealers to form trusts, cartels, or "rings" to control price and supply. Indeed, by the late 1880s, buyers' efforts to control the market had become endemic and were seen by outsiders as "old hat." According to the *Boston Globe* in 1888, "Having found their strength, the fish dealers have for the last 10 years been continually engaged in forming rings." In this particularly audacious attempt, all but one of the fish dealers in Boston and New York established a single association in 1882 to control northeastern fish markets. They rented offices, stores, and warehouses; hired clerks and accountants; designated a clearinghouse for select buyers; and elected officers and treasurers under the name of the Massachusetts Fish Exchange. Still, the one firm that refused to join, the Bay State Halibut Company, challenged the ring. That firm, instead, bled the Massachusetts Fish Exchange dry by paying higher prices to fishermen and selling products at a loss. Disrupted deliveries from the Fish Exchange further compelled retailers to turn to Bay State for a more dependable supply. One by one, Massachusetts Fish Exchange members filled their own orders from Bay State, and the trust ultimately collapsed.[3]

As the Bay State Halibut Company demonstrated, controlling the fresh fish market was not easy. The product's short shelf life forced would-be monopolists to move beyond conventional patterns of vertical or horizontal integration. Bad fish and irate buyers posed challenges too great for conventional tactics involving rebates, boycotts, and control of raw materials to be successful. Still, with the vast majority of fresh fish landed in Boston for distribution across the eastern seaboard, market control could be accomplished by controlling the supply chain at the critical point where fish went from boat to buyer. By the mid-1880s, Boston dealers understood that whoever physically controlled the docks where fish came ashore also effectively controlled the northeastern market, and perhaps the supply of fish all along the seaboard.

Despite efforts to enforce a profitable stability on highly variable fish markets, wild economic swings continued to rock the national economy through the 1880s. It was only a matter of time before Boston fish dealers again tried to form a trust to bring some order to their industry. The most important step toward that goal came in 1885, when a consortium of dealers struck an agreement with the city to concentrate all fishing activity onto Boston's T-Wharf in the heart of the waterfront (which has since been swallowed by development). From a public perspective, this was a good idea: the deal removed the noise, smell, and congestion of the fish business from the rest of the waterfront.[4] It proved far more salutary for the dealers, however, because it concentrated all fishing activity at a single pier, forcing fishermen who wanted to sell to the Boston market to offload on the pier they could more easily control.

Consolidation onto T-Wharf took an ominous turn when, three years later, forty-five fish wholesalers on the pier formed a cartel to remove what they saw as needless competition, a move clearly in line with the business culture of the time. Combining as the Boston Fresh Fish Dealer's Association, the cartel established its headquarters on T-Wharf and launched its gambit to control prices for fresh fish throughout the Northeast. With nowhere else in the city to land their fish, fishermen had to accept what association dealers offered. Retailers faced a similar bind: with no other supply available, they had to pay whatever the cartel demanded. For nine months, the Fresh Fish Dealer's Association had both fishermen and consumers over a barrel. Trust hardly defined the cartel, however. Fearing that some members were selling outside the organization, cartel officers hired detectives, who found that a dozen members were "not satisfied with their legitimate profits." Under fictitious names, they sold fish below the cartel's set price. Soon, the ring exploded. In its aftermath, wholesalers relentlessly slashed prices until, as the *New York Times* put it, "many of the smaller dealers who were protected by the trust, [were] forced to the wall," and "bankruptcy thins the number."[5]

Dealers in other ports tried to control the market as well. As the Boston Fresh Fish Dealer's Association was imploding, a Gloucester cartel that included large dealers in New York and Boston "flourished" by controlling the halibut market for two years in a row.[6] Three years later, newspapers reported another "gigantic fish trust" breaking apart.[7] With rumors of yet another new trust filling the air, one witty journalist penned: "Now all the fishes of the ocean / Are today in great commotion / For the fishers think they must / Straightaway form a cod-fish trust." Newspapers reported two more attempts in 1894 and 1895: in the former, Gloucester dealers wanted to counter the "ruinous character" of their competitiveness.[8]

While many dealers tried to corner the fish markets, only those in control of the docks had any chance to succeed. In 1908, the same year that the Gorton-Pew Fisheries Company formed in Gloucester, T-Wharf dealers made another attempt at combination. In forming the New England Fish Exchange, they institutionalized the disproportionate power they enjoyed over fishermen. In order to sell on T-Wharf, fishing captains had to accept the exchange's new terms, laid out in its formal "Guarantees and Agreements for Control of the Fish Business." Under these terms, captains had to agree to sell fish only to the exchange, in rooms designated and used exclusively by members of the exchange. Furthermore, they had to pay a 1-percent fee on the value of their catch "to help defray the expenses of this Exchange." Landing fees quickly exceeded operating costs, however, and the surplus was deposited in a pool and doled out to members as a dividend.[9] Under this business model, the New England Fish Exchange became one of the most powerful forces in New England fisheries. It soon challenged Gloucester, confronting the outfitters' carefully crafted image of what New England's fisheries were supposed to look like.

Resistance to Boston Dealer Dominance

Dealers throughout the region were increasingly frustrated with Boston's control over the fish market. In 1896, a New York merchant decided to break Boston's power and consolidate the entire fishing industry in his city. According to the *Boston Journal*, this dealer declared, "The Boston commission merchants and fish men for a long time have had practically a monopoly . . . while they themselves do not own any fishing craft, and, in fact, many of them have not a dollar invested in the industry, but only sell the commission at a large profit while the producers very often do not get sufficient returns to pay the expense of the voyage."[10] He was unsuccessful. Another dealer attempted to get around Boston's monopoly by opening a large pier in Beverly, Massachusetts, in 1898. This plan also failed, as did three more attempts to break the trust, reported in Boston papers over the next two years.[11]

Some fishermen pushed back as the fish dealers consolidated their power, but their ire flew in the wrong direction. This was not surprising given the industry's complexity—fishermen, captains, vessel owners, fish buyers, and lumpers all took cuts from the profits of a voyage. Regardless, in 1886, fishermen struck against the vessel owners and won a greater share of the trip profits. The owners also agreed to pay more of the cost of ice and bait. Yet despite these concessions, which were an important victory for organized labor, the underlying problem with fishermen's earnings remained. Vessel owners drove hard bargains, but behind them, T-Wharf dealers drove even harder ones. Colluding to depress the price paid for fish, Boston's fish dealers, even more than vessel owners, undermined fishermen's pay.

Five years later, Boston's haddock fishermen, now members of the Deep Sea Assembly of the Knights of Labor, struck over the same issue. Crews on two vessels tied up at T-Wharf and refused to cast off until they got a better deal. The next day, fishermen laid up two more schooners as more men joined the union and the protest. Four vessels chained to the dock deprived Boston's markets of 150,000 pounds (56 mt) of haddock a day, so most vessel owners again struck a deal.[12] Yet the underlying issue remained. With all of Boston's fish coming through T-Wharf, dealers wielded tremendous power over price, which no strike against the owners could address.

Clearly, no free and competitive market served Boston's fish consumers during this period. Persistent attempts to corner the fish market, made possible by the concentration of fishing businesses on T-Wharf, distorted prices and drove down pay despite competition among producers. As in Gloucester, the scrum to gain advantage infused chaos into an already unpredictable industry. If a consortium could control Boston's fish pier, however, they could control markets across New England and the eastern seaboard. While Gloucester firms sought to control the means of catching fish, Boston dealers controlled the bottleneck, the portal where fish were landed and processed, and out of which they were distributed. By 1908, few could tell which port would emerge victorious.

The Popular Appeal of Fisheries in an Age of Industrial Chaos

Persistent press coverage detailed another important issue in these political and commercial struggles. News stories reflected intense public interest in New England fisheries' politics and practices. Whereas before 1885, when fisheries debates were sequestered in the pages of dry political journals, business papers, and administrative records, by the mid-1890s, stories consistently appeared in the region's largest newspapers. This attention stemmed, in part, from the commercial success of Rudyard Kipling's *Captains Courageous* (1897).[13] With the broad cultural discourse over New England's maritime heritage in an industrial age, the chronic, politically symbolic debates over the free entry of Canadian fish into American markets fueled popular interest in fisheries issues. Through the 1880s, Gloucester's position had dominated discussions. As Boston's influence grew during the next decade, however, its dealers began to actively oppose Gloucester's positions in order to secure cheaper Canadian fish. Boston's largest newspaper, the *Boston Globe*, and other popular publications regularly reported the different stands Gloucester outfitters and Boston dealers took on tariff policies and treaties. Inflammatory headlines such as "Boston Amused... Deny Gloucester Has Any Cause for Indignations," "Stupid and Suicidal," "No Use for It," "From Fishers to Men-of-War," and "How New England Was Threatened" attracted a steady readership for the tit-for-tat commentary of competing fishing interests. What began in 1885 as a one-time gambit to win over public opinion had become routine operating procedure a decade later.[14]

Most mesmerizing, however, was coverage of the hazards inherent in fishing: sensational at-sea rescues, shipwrecks, drownings, and other calamities that routinely struck the New England fishing fleet. Press accounts of miraculous rescues and heroic survival attracted widespread public attention and sold more papers. In November 1899, for example, the *Boston Globe* printed an extra edition when the schooner *Joseph P. Johnson* rescued the crew of the *Alfaretta S. Snare*, who for two days watched seas break over their vessel as it foundered seventy miles east of Cape Cod.[15] Fears for fishing vessels returning late routinely made headlines, too, as did the rare occasions when no deaths or injuries took place. "Not One Man Lost," declared one story, and "Remarkable Immunity of Provincetown Fishermen from Disaster during Past 12 Months" topped another.[16] Boston newspapers between 1897 and 1904 ran at least twelve stories detailing fears for lost vessels, their fortunate or heroic returns, or their tragic ends.[17] Headlined with bold language, Boston newspaper kept readers informed of the plight of the fisheries. Recently reminded of their fisheries' role in national culture, Boston readers thrilled to the exploits of their beloved industry. It became almost a moral mandate, persistent coverage suggested, to know the human costs of seafood: "Seven Days on the Banks: Lives Risked That Bostonians May Enjoy Cod."[18]

With so much interest in New England fisheries, savvy business leaders continued to use the press as a political tool. Individual fishermen did the same thing. One captain in 1899 invited a journalist from the *Boston Globe* to sail aboard his schooner for a month-long trip to the Nova Scotian banks. The resulting human-interest piece, headlined "Catch of 10,000 Pounds Made on the First Day," brought to readers the day-to-day challenges of tub-trawling (catching groundfish with baited hooks, lines, and dories) on distant water banks as it celebrated the continued success of the schooner fleet.[19] Other stories were puff-pieces, pure and simple. They illustrated nonetheless the depth of popular interest in the region's fisheries: "Regular Tar, Every Inch of Him" described Leo the dog, mascot of the schooner *I. J. Merritt* of Provincetown.[20] The 1904 story "Maine's Fisherman Preacher" recounted the toil and ministry of an eastern Maine fisherman tending to the spiritual needs of island communities. "Proud Father of Nine, He Relies on the Sea for a Living, and Spreads the Gospel Gratuitously—Picturesque Life of a Striking New England Character," the headline boldly declared.[21] Human-interest tales like these went beyond fisheries and port politics. They responded to national curiosity about what was, arguably, New England's oldest industry, and one that had lately come to define its character.

Even the U.S. Bureau of Fisheries (USBF), the federal agency tasked with fisheries research and commercial development, made use of the popular press. In 1912, the service published Francis Rolt-Wheeler's *The Boy with the U.S. Fisheries,* a novel about a boy finding himself (conveniently, from the services' perspective) in some of the most adventurous and exotic places where the USBF undertook field research. In the preface, Rolt-Wheeler unequivocally celebrated the nation's fishermen. Just like Dr. Edwin Gould's characterizations two decades earlier, Rolt-Wheeler believed that fishermen maintained a boldness all but lost to the past and embraced the very values needed to sustain it in the future: "Treasure-ships, bearing richer cargoes than any galleon that crossed the Spanish Main, still sail over the ocean to-day, but we call them fishing smacks; heroism equal to that of any pioneer navigators of old still is found beneath oilskins and sou'westers, but the heroes give their lives to gain food for the world instead of knowledge." Rolt-Wheeler also added a modern twist. Along with fishermen reaping the sea's harvests, so too must Americans celebrate scientists working to understand its secrets. Indeed, the scientists working within the USBF were an antidote to war and violence: "Just as to save life is greater than to destroy it, so is the true savior of the seas the [U.S.] Fisheries [Service] craft, not the battleship; so is the hatchery mightier than the fortress, the net or the microscope a more powerful weapon for good than the torpedo or the Nordefeldt [anti-torpedo gun]."[22] Richly illustrated from the bureau's own photographs and engravings, Rolt-Wheeler sent his young protagonist around the world with USBF researchers in an odd pastiche of boy's adventure literature, heroic science, imperialism, and nationalism.

While the stories discussed above helped cement the iconic fisherman in New England culture and galvanized public opinion in fisheries debates, no portrayal of fishermen reached the prominence attained by Rudyard Kipling's *Captains Courageous*. More than any other turn-of-the-century representation of New England fishing, Kipling's popular book embedded in public consciousness an image of the heroic and noble fisherman whose virtues and bravery posited an antidote to contemporary social change. Kipling's story is familiar enough: the well-to-do wastrel teenager Harvey Cheyne falls overboard from a transatlantic steamer while on a trip with his rich industrialist parents. Cheyne is miraculously saved from certain death by a fisherman, Manuel, working trawls from the Gloucester schooner *We're Here*. After Cheyne accepts the reality that the ship's Captain Disko Troop feels scorn for his obnoxious, elitist, disrespectful new crew member, a process begins by which fishing, the crew, and the vessel itself transform Cheyne into a hardworking, thrifty, productive young man.

Kipling latched onto the story in 1896, learned from his Vermont neighbor James Conland, who had worked in the banks cod fishery in the 1860s. Shortly thereafter, Kipling visited the fish piers in Gloucester and Boston, taking in as much detail as possible. Despite seeing the fishery as a vibrant, ongoing concern, Kipling interpreted his visit through a veil of decline.[23] He wrote in *Something of Myself* (1937), "I wanted to see if I could catch and hold something that was already beginning to fade."[24] For Kipling, the New England fishing industry appeared the perfect foil to contrast the old America of hard work, bravery, honesty, and simple genuineness with the new industrial America driven by profit, production, and speed.[25] Kipling's textual changes during the preparation of the book reveal his desire to idealize social relationships among the crew. For example, Kipling edited out the more aggressive taunts his characters issued to passing vessels but left the good-natured ones in to show a rough but benign edge to fishermen's camaraderie. He also reduced the intensity of the conflict between two crewmembers, Salters and Penn. As the Kipling scholar Leonee Ormond notes, "Other omissions were made to lessen any impression that the crew of the *We're Here* might be rough or brutal."[26] Like observers before him, Kipling idealized the men aboard the *We're Here* and portrayed New England fishermen not as individuals but as icons in this passing way of life.

Kipling's handling of the manuscript had consequences, however. Literary critics in both the United States and Britain found the story less than compelling, due in part to its idealized characters. For example, an *Atlantic Monthly* reviewer stated, "Most of the characters with the exception of Disko Troop are mere outlines, distinguishable by Dickens-like tags."[27] Regardless of its critical reception, the book sold well in the United States, with its conservative morals, its simple allegorical message, and its homage to days gone by. By the time Kipling died in 1936, *Captains Courageous*, although not viewed as one of his best literary works, was one of the most popular.[28] That popularity cemented into the public mind at the turn of the century a saltwater world filled with courageous

New Englanders navigating both seas and markets with honesty, integrity, and moral virtue.

If in the early 1890s people saw New England's fishermen as embodying the virtues of American independence and hard work, by the end of the decade fishermen were also seen as preserving the ethnic purity of America's British roots. Public coverage of the Canadian tariff deliberations reveal that for many observers, Yankee fishermen embodied the labor discipline that conservative Americans valued. They also served as a bastion of Anglo-Saxon "stock," which nativists feared increasing immigration threatened. In fact, New England's fisheries had long employed anyone willing to work on its terms, regardless of ethnicity. For example, Native Americans played an important, if exploited, role in the whale fishery in the eighteenth century, and a significant number of Portuguese fishermen exercised considerable influence in the banks fisheries, especially in the high-stakes mackerel fishery, as early as the 1840s. Furthermore, throughout the nineteenth century, "white-washed Yankees," Nova Scotians not necessarily of British descent, settled in New England after working aboard American fishing schooners.[29] Since the 1870s, however, reorganization of the New England fisheries helped fuel nativist fears of Anglo-American decline. Following the Civil War, vessel owners and captains began moving their operations to larger ports that attracted foreign fishermen in great numbers.

For middle-class American travelers enjoying seaside vacations in coastal New England towns, quiet fishing wharves and the presence of foreign-born fishermen evidenced a decline and dilution of New England's Anglo-Saxon cultural values. Immigration from southern and eastern Europe peaked in the 1890s. While the proportion of immigrants to native-born Americans remained consistent in cities, white elites began to develop racialist arguments against immigration that contrasted the newcomers with mythically democratic, Anglo-Saxon, western European descendants. Despite their obvious historical diversity, New England fishermen and fisheries seemed to personify the very virtues and values that new immigrants threatened. Many middle-class vacationers, arriving in summer in New England fishing towns, perceived ethnic diversity not as a continuum of historical trends but as a recent deterioration of cultural security.[30]

Thus, ethnic dilution and structural change made many fear for the future of New England's fisheries and the social consequences of the industry's demise. In February 1896, the popular yachting writer and historian of the America's Cup races Winfield Thompson penned an article bluntly entitled "The Passing of the New England Fisherman" for *New England Magazine*: "Go down to the New England coast, wherever you will from Nantucket to Quoddy Head, and you will witness the passing of the New England deep sea fisherman. Not many years ago he was on a practical equality in respect to the importance of his craft, with the farmer, and, with his sturdy qualities, he was a force among his fellows." Like Edwin Gould, Thompson saw this old-guard mariner as a complacent worker: "He saved what he could from his earnings after all the mouths dependent upon him were fed, and paid his taxes without questioning the theory which might underlie the system by which he was taxed." Despite strikes among Boston halibut fishermen in 1886 and in 1892, Thompson pushed his quixotic visions further, claiming that because of the shares system, American fishing "always has been and is now free from dissension and anything like 'labor' troubles. . . . The fisherman . . . is satisfied with what he gets, providing Fortune favors him in his catch." He was "no dullard, this Yankee fisherman, and he was no visionary. He went fishing and braved danger because there were a good many little ones to feed and clothe at home, and nobody to do it but himself." In sum, "the change in blood and personnel in the New England fisheries becomes interesting mainly from the economic and pathetic features it presents, in the decay of coast villages and the disappearance of a sturdy type of American citizen." Concentration into a few central ports marginalized "old stock" Yankee fishermen in their small port communities. With new blood entering the fishery, American society, prosperity, and its work ethic were endangered: "With this concentration of the fishing business at a few ports, a new type of fisherman, of foreign birth, less citizen and more laborer than his predecessor, has come in; and the small villages, deprived of their main support, have retrograded."[31]

With that retrogression, Thompson envisioned an important center of civic values falling into decay, like the rotting wharves and derelict fish houses of the small fishing ports. In contrast to the "the typical New England fisherman [who] was a sturdy, wholesome citizen, as hardy as good blood and tough labor could make him, clean-minded, blunt in speech, open-handed, generous and confiding," came the "thrifty sons of England's North American possessions the blonde and frugal Swede and the swarthy Portuguese" who sent their earnings overseas instead of spending them in the United States. Immigrant fishermen, Thompson believed, undermined Yankee values, communities, and the economic health of the region. "If the New Englanders feel any resentment against the new men who have secured a footing in the fisheries, it is perhaps directed more toward their Provincial [Canadian Maritime] cousins, who make their money here and spend a large part of it at home, than against the new comers from beyond the Atlantic who become citizens and spend their money where they make it." Regardless of who was to blame, Thompson mourned what he interpreted as the loss of a group of people who, he felt, preserved core American values. As Yankee fishermen left the sea and found better paying jobs in factories or built farms further west, America lost the ideal fisherman who "tried to rear his children to become as honest in purpose as himself."[32]

The fact that Thompson's views appeared in a popular literary magazine suggest wide acceptance. Whether it was nativism or merely benign yearning for a fictive golden age, Thompson and other tourist writers appealed openly to their readers' fear of lost racial purity.

Mechanization, Monopoly, and the Establishment of New England Steam Trawling

Even as the public embraced New England's fishing image of continuity and character, technological innovation in Boston was about to change fishing forever. In November 1906, front-page stories in both the *Boston Globe* and the *Boston Journal* reported that a "Gigantic Fish Trust" was forming on T-Wharf. Capitalized at over $5 million and supported by more than two-thirds of the wharf's fish dealers, the syndicate planned to consolidate operations, thoroughly mechanize the fishing process, and bury Gloucester firms once and for all: "It is proposed to establish a large fish curing and drying plant, and . . . with such establishments in operation here, it is believed that a large percentage of the fish-curing business now controlled by Gloucester would be brought to this city."[33]

In its reorganization, the trust would dispense with the theatrical fish auctions that had defined T-Wharf since 1885. Instead, it would set standard prices ostensibly to stabilize the market by removing what one account dubbed "senseless competition." If successful, the cartel would "revolutionize the fishing business of Boston and make this city the one and only great fresh fish mart of America."[34] The trust, called the National Fisheries Company, bet its future on the only steam-powered trawlers then operating in the United States. One member of the trust, the Bay State Fishing Company, had developed a method to drag a large net through deep water to scoop up fish. Launched in 1905, their first beam trawler, the F/V *Spray*, pulled a conical net held open by a long beam—a beam trawl—over the bottom in pursuit of groundfish. In Europe, sailing vessels had used nets like *Spray*'s sporadically since the late Middle Ages. Steam-powered vessels, a British innovation, had come in the 1860s.[35]

Despite this pedigree, New Englanders approached beam trawling tentatively, and the gear's arrival generated great controversy. In 1864, Richardson Leonard built the beam-trawling schooner *Sylph* specifically to catch halibut, the largest bottom-dwelling flatfish. Most halibut eluded the trawl, but those that did not came to market too mangled to fetch good prices. More importantly, Leonard's vessel and crew attracted heated opposition from Irish hook fishermen in Boston, who blamed trawls for destroying small-scale fishing in the United Kingdom. These fishermen did not intend to see this happen again. According to the *Boston Globe* in 1891, "As soon as war was declared against the [beam trawling] vessel and crew . . . they were forced to take the [trawl] out to sea and dump it overboard with all its paraphernalia."[36]

J. W. Collins saw potential for American fishermen, however, and published the plans of British steam-powered beam trawlers in the 1889 *Bulletin of the United States Fish Commission*.[37] Two years later, Captain Alfred Bradford reported that his vessel, the *Mary F. Chisholm* of Boston, successfully trawled for haddock. By the end of the 1891, the British beam trawler *Resolute* was fishing in the northwest Atlantic and offloading fish in Boston as well. The Boston Fish Bureau, the information service of Boston fish dealers, saw great opportunities in beam trawling. Highlighting the gear's ability to catch fish previously unknown to hook fishermen, the bureau's secretary exclaimed, "Such a variety of fish was never known to arrive at the port by a single vessel at any other time." Trawling promised to transform the flatfish fisheries altogether. Flounders and flukes were not injured by the trawls, and they had never before "been caught in large [enough] numbers" to supply a substantial market. "This method of fishing will, therefore, introduce in our markets new species of flatfish which have heretofore been practically strangers."[38]

Yet some Boston dealers remained skeptical. They complained that the fish were too mangled by the gear to fetch good prices. Other dealers "laughed at the idea of Gloucester men going to about $20,000 expense to build and fit a vessel to catch [low-valued] flounders and two-pound haddock, which was about half her cargo." Haddock fishermen, who had unionized and gone on strike in August 1891, also had doubts. To them, the *Resolute* and the *Mary F. Chisholm* represented the owners' counterthrust to fishermen's unionization. They also believed that beam trawling threatened their livelihood in several ways. Beam trawlers caught more fish with fewer hands than tub trawling or hand lining could catch. Furthermore, fishermen feared that the capital-intensive nature of building, outfitting, and operating the steam vessels made beam trawling fisheries unlikely to adopt the shared compensation system, despite its shortcomings, commonly found on schooners. The gear posed additional challenges to the fisheries as well. One fisherman who had worked beam trawls in the English Channel and Irish Sea told the *Boston Globe*, "Leaving all jokes aside . . . should this trawl prove a success in this country, a fish famine will surely follow, as the foot of the trawl, which scrapes along the bottom of the sea, not only destroys the feed for fish, but also destroys the spawn."[39]

Fifteen years later, in November 1906, these fears were justified when a group of Boston dealers, outside investors, and the Fore River Ship and Engine Company formed the New England Fishing Company. The amalgamated firm included, among others, the Bay State Fishing Company, which operated the only beam trawler in New England, and the New England Fish and Halibut Company, which operated three beam trawlers in the Pacific Northwest. Although this "Gigantic Fish Trust" initially paid fishermen on shares, observers acknowledged that the firm sought to profit from trawling's high volume catch rates and low labor needs. More fish landed by fewer fishermen could only serve Boston's dealers.[40]

Romanticism and the Resistance to Steam Trawling

Fishermen were not alone in making this observation. The press immediately spelled out for their readers the changes looming ahead for New England fishing. The *Boston Journal*'s headline "Steam Trawls Spell Doom of Fishermen . . . Passing of the Hand Liners" prophesied likely pay cuts for fishermen and ruin for schooner owners.[41] There was good reason for alarm. T-Wharf dealers saw sailing vessels as relics of an unconsolidated, unstable, unprofitable, and outmoded past that could vanish with the ascendance of beam trawling. The New England Fishing Company deliberately excluded sailing vessels from its operation, requiring that members that owned sailing vessels to manage them independent of the cartel.[42]

Virtual battle lines formed instantly between Gloucester's venerable schooner fleet and Boston's new steam trawlers. Even before the trust was formed, Gloucester captains had considered petitioning Congress to ban the use of beam trawls in fishing. Extemporizing the best argument they could come up with in a pinch, the captains declared that "the work of the government in spending thousands of dollars for the fish hatcheries at Woods Hole is largely undone by the beam trawling, for while the old hand lines take only the larger fish, and allow small ones to escape, the beam trawls scoop in everything." Against this mostly ecological claim, the trust's supporters countered that beam trawls had "revolutionized the fishing business" in England, where fishermen sold their sailing outfits to Sweden and Norway. Furthermore, *Spray*'s impressive landings, after two disappointing startup months, made a clear economic argument: "On the strength of this showing, capitalists have been urged to buy stock in the new enterprise, because it would pay 6 percent."[43]

Still, the powerful New England Fishing Company and its new technology did not spell doom for the sailing fleet right away. As New England fisheries trusts had formed and collapsed with brutal regularity over the past three decades, schooner owners adopted a wait-and-see approach, seeking redress instead for the off-loading, wharfage, and ice fees they were forced to pay at T-Wharf.[44] Ironically, Boston's turn to steam trawling may have inspired the formation of the Gorton-Pew

Fisheries Company, as Gloucester's remaining outfitters sought protection and business security by consolidating ownership of the city's remaining schooners.[45]

While few saw it coming at first, over the next six years Gloucester and Boston positioned themselves for a confrontation that not only shaped the future of regional and national fisheries but relied even more heavily on politicized representations of New England fishermen. In 1912, Nova Scotia banned beam trawling and its more efficient derivation, otter trawling. Instead of a beam, this new gear used large doors called "otter boards," which were angled to hold open a much larger net as it was towed through the water and dragged across the ocean floor.[46]

Inspired by this precedent, Gloucester outfitters pressed their congressman, Augustus Gardner, to put similar legislation before the House Committee on Merchant Marine and Fisheries. Dubbed the "Gardner bill," the proposed legislation would effectively ban the landing of fish caught with beam and otter trawls, and in essence the gear itself. Two different visions of the fisheries quickly emerged in the ensuing committee debate. Gardner, like Gould before him, drew on the vision of New England fisherman that Collins and others had launched and that had evolved in the popular press over the past three decades. For Gardner and his Cape Ann constituents, the fisheries were about tradition, heritage, manly virtue, and American social development. His opponent, William F. Garcelon, lawyer for the Bay State Fishing Company, argued that the fisheries were a modern industry: innovation, mechanization, capitalization, efficiency, and safety (in both the workplace and the final product) should guide fisheries management. In short, the bill pitted Gloucester's romantic ideal of New England's fishing heritage against the industrial reality of Boston's modern fisheries.

In fact, efficiency was the biggest argument Gardner had to overcome in defending his proposed *de facto* ban on otter trawls. Fishing with hand lines and tub trawls simply could not catch the quantities of fish hauled up by steam-powered beam and otter trawls. To win the debate against raw economic statistics, Gardner presented an emotional, qualitative argument. First, he used motion pictures to show the House committee film footage of mutilated fish hauled aboard a steam trawler. Because they were too mangled for market, much of those fish were thrown back overboard, dead. Second, he brought in British fishermen and American steam trawler officers to testify about the gear's destructive effects on fish stocks and on the fishing communities themselves. Visual evidence of waste and testimony from eyewitnesses to this destructive new form of fishing gave Gardner a strong moral foundation for his case.

To build a stronger case, however, Gardner needed to show the broad effects this form of fishing would have on the wider New England society. To do this he called an important ally, Gloucester journalist and novelist James Brendan Connolly, whose research into European steam trawling a decade earlier presented the new gear in a grim light. Equally important to Gardner was Connolly's later career. Since 1904, Connolly's novels had rehashed Kipling's stalwart fishermen motif. During eight years, he had churned out eight novels, including *The Seiners* (1904), *The Deep Sea's Toll* (1905), *Out of Gloucester* (1906), *Crested Seas* (1907), and *Open Water* (1910); each tells the fictionalized exploits of brave Gloucester fishermen. The American reading public embraced the author's work warmly, and the *New York Tribune* added, "No more adventurous set of men live to-day than the hardy fishermen of the Gloucester fleet, and no one has caught the spirit of their life better than *James B. Connolly*."[47] For Gardner, he was the perfect person to speak to steam trawling's wastefulness and to the social benefits of hook fishing. With so many works enjoying such popularity, it would be difficult for the committee not to think of Connolly's heroic fishermen as he recounted his European findings. By placing the journalist/novelist on the stand, Gardner sought to invoke the symbolic imagery of the New England fisherman that Connolly had crafted so well and which, once before, had appealed to the committee's head and heart.[48]

But this was a new day. In the committee hearings, Garcelon's strategy crushed Gardner's tactics. The Bay State Fishing Company lawyer attacked Connolly's claims of expertise. Garcelon's questions exposed the outdated nature of Connolly's work, its paucity of reliable statistics, and its reliance on personal impressions over hard data. After establishing that Connolly knew little about fish behavior, morphological differences, and North Sea fisheries statistics, Garcelon branded Connolly an expert on adventure stories and expounded on the impropriety of a novelist testifying to the committee. That was the point: Garcelon used his questioning to redirect the committee's attention away from the romanticized images Gardner and Connolly were presenting and turned it toward substantive questions about what was best for the fishermen. Instead of romance, Garcelon presented a new vision of fishing based on modern notions of efficiency, safe working environments, wholesome products, and profitability. His point was clear. Did the committee wish to continue sending virtuous, hardworking fishermen off to die in fogs and storms for the sake of lost visions of New England's past? Or did they want to support a sophisticated, modern industry that used state-of-the-art technology to reduce risk and increase catches for the good of investors, fishermen, and consumers? Garcelon argued that if Gardner's bill passed for the sake of sentiment, it would prevent New England fishermen from enjoying better pay, safer working conditions, and more regular work schedules, and instead perpetuate an industry many saw as dangerous, inefficient, unpredictable, and unsanitary. In short, Garcelon brought into the open the dangers of discussing fisheries policy in terms of romanticism, which had enthralled participants in fisheries debates for more than thirty years.[49]

In the face of Garcelon's argument, Gardner needed statistical evidence to support his case. To get this, and to buy time, he called for the USBF to investigate the destructiveness of otter trawling. Sending the issue off for additional research would delay a final vote on the bill and provide the committee with hard data on how wasteful otter trawling really was. Garcelon agreed to this proposal, likely because the study would also show how much fish the gear could really catch. Carried out in 1913, the final report confirmed the gear's wastefulness: roughly two-thirds of the fish landed had to be thrown overboard dead, too mangled for market. Yet the report also confirmed the gear's efficiency: even with the waste, steam trawling landed far more fish, of all sizes and greater variety, than hand lining or tub trawling. In the face of such massive productivity, and with little concern for waste in what many considered to be an inexhaustible resource, the committee killed Gardner's bill.[50]

In the end, this was a struggle over who would control the image of the New England fisherman: the people who helped create the heroic image of the fisheries or the people who were trying to improve their operational safety and profitability. Heroic imagery had helped to defeat the Lapham bill two decades earlier, when notions of humble, virtuous fishermen swayed leaders at the center of government. But Gardner stretched the permeable boundaries between romance and reality too far. By presenting a novelist as an expert witness in a hearing about the modern fishing industry, he revealed how deluded he and his Gloucester constituents were about the future of New England fisheries. A smart lawyer, Garcelon turned Gardner's gambit into a parliamentary fiasco, highlighting the disconnect between romantic representation and reality. Trawling had arrived in New England to stay.

War Profiteering and Corruption in Boston's Trawler Fisheries

Following the Gardner bill's defeat, the Boston fishing industry expanded its hold over New England markets by investing in steam-powered otter trawlers to target groundfish, particularly haddock. Dockside control allowed a few individual firms to dominate the region's supply of cheap protein. In 1908, the Bay State Fishing Company, the New England Fish Exchange, and other firms interested in the fisheries business organized the Boston Fish Market Corporation (BFMC) through a series of interrelated companies. This firm, although technically a consortium of different firms, was controlled by a small number of people through interlocking directorates. Through this network, the new cartel affected almost complete control over the fresh fish market in Boston, excluding nonmembers from buying fish in the city and levying surcharges on fishermen landing cargoes on T-Wharf.[51] With the funds it amassed, ostensibly to advance the fishing industry as a whole, the cartel inflated dividends paid to shareholders. They also leveraged a hefty public subsidy for the 1914 construction and leasing of a new, modern, purpose-designed and -built pier situated across Fort Point Channel from downtown Boston. Dubbed the Boston Fish Pier, the new facility was dedicated to serving the fish buyers' and fishermen's needs. The cartel used surplus funds to expand its holdings via kickbacks and generous payments to pressure competitors to sell out. By 1917, between the levy on fishermen and the profits gained through consolidation, BFMC shareholders secured extraordinary profits of 32 percent per share just on the less remunerative common stock. World War I provided the combination with additional opportunities to make hefty profits. As the war drove up food prices across the Atlantic in 1914 and 1915, the cartel continued to consolidate its position. By 1917, they had used wartime exigencies to drive up the average price of cod from $3.12 in 1915 to $4.04 in 1916, and to $5.05 in 1917.[52]

Whereas "pay to play" fees and rigged auctions had defined Boston's fresh fish market since 1885, inflating wartime food prices went too far. By February 1918, the dramatic price increase in some of the cheapest protein available to working American families caught the attention of state legislators. Beginning in March, the Massachusetts General Court created a joint special committee to investigate claims of wartime profiteering and violations of federal antitrust laws. The committee's initial findings supported the subsequent creation of a joint special recess committee to investigate corporate price-fixing across all Massachusetts fisheries markets, including Boston's fresh and Gloucester's salt fish markets. As a result of the legislative investigation, which subjected Boston's fish cartel to some of the most intense legal scrutiny in the history of New England fisheries, a slew of criminal cases unfolded as the state sought to break up the cartel and punish the responsible parties for illegal market activities.[53]

As the investigations commenced at the height of the war and targeted a group of alleged war profiteers, they attracted immediate, widespread media attention. During the two years of hearings, newspapers ran daily stories detailing the most recent revelations. The press paid even closer attention to subsequent criminal trials, convictions, and sentencing. Often, coverage made the front page and rarely was it sympathetic to the defendants. Front page headlines such as "Government Opens War on Boston 'Fish Trust,'" "Boston Ex-Mayor Attacks Fish Trust," and "Indict 30 on Fish Monopoly Charge: President Dyer, Financiers and Men Prominent in Trade Named" helped fuel an almost ghoulish, citywide hunger for news about the cartel's legal woes.[54] In contrast to dull legislative and legal proceedings tackling obscure regulatory or tariff provisions that few cared about, the state inquest into the cartel's manipulation of the market, consumers, retailers, and fishermen fascinated and enraged people throughout the Northeast. Far from a link to New England's virtuous past, Boston's fisheries proved to be all too modern in their business dealings.

The committees' findings validated state suspicions that the BFMC had rigged the market against consumers during a national wartime emergency. In presenting its conclusions, however, the joint special committee confronted not only a corrupt and powerful suite of firms and an outraged public but the world at war: "In the present [wartime] emergency the Committee feels that with the eyes of the people turned toward their departing sons, with the attendant strain of war, taxation, transportation and labor problems, with the burdens occasioned by the high cost of living, it would be unfortunate if anything should be done by the Commonwealth to disrupt or hamper the production of necessaries of life, but steps should be taken to safeguard and protect the consuming public from profiteering and unfair practices." In short, the committee balked at punishing the transgressors. "Bearing in mind that the fish industry is one of two industries left to Massachusetts, we should be slow by hostile acts or legislation to so regulate or legislate against it in this State as to drive it to competing centers."[55] With that in mind, the committee presented a series of recommendations to address the problems the scandal revealed. First, and curiously beneficial to the guilty parties, the committee requested that the federal government return the steam trawlers the U.S. Navy had requisitioned into service in 1917. The committee additionally recommended that all dealers willing to pay a reasonable fee be allowed access to fish auctions and that offloading fees charged to fishing vessels be reduced.

Most important, however, was the committees' determination that all vessels should put their fish up for auction. They found that the cartel had been "absorbing large quantities of its own fish arriving at the Pier through its distributing branches, and not selling them through the Exchange, but still bidding upon the fish of other producers on the Exchange." Doing so allowed the combination to sell its fish "at a price less than it forces its competitors on the Exchange to pay for fish obtained by them." Three outcomes ensued. First, the cartel could sell its own trawlers' catches at higher prices; second, it "unwarrantably" drove up the prices external buyers paid for fish on the exchange; and third, using the high volumes of fish caught by its own trawlers, the cartel could undersell its competitors and eventually force them out of business. It was to obviate this last issue that the committee required all bidders to put their own fish up for auction if they wished to purchase additional fish. The BFMC could no longer use its constituent's trawler fleets to rig the market. Like all outside firms and operators, members of the cartel had to sell all their fish at open, public auction—they even had to buy back the fish caught on their own trawlers.[56] In effectively erasing the differences between larger and smaller vessels in the market, the court laid the foundation for New England's persistent small-boat fishery, reshaping fisheries for decades to come.

If the joint special committee sought to avoid conflict with Boston's corrupt fish dealers, the recess committee did not. It not only accepted the first committee's findings; it went further, calling to task many more who profited in the scheme. "In the opinion of the Committee, the cause of high prices cannot be fixed upon any one group who have a part in the distribution, but the fisherman, the wholesaler and the retailer are in part responsible for the advance in price, and all have profited at the expense of the consumer." Neither a call to return to past ways nor an indictment of Boston's modern trawler fleet, the committee adopted a middle-of-the-road approach. In the case of the fishermen, the trawler fleet's reliance on fixed wages marked an improvement over the traditional share system. The share system, the committee found, clearly yielded inadequate compensation given the hardships and dangers involved in schooner fishing, yet "the profits which are shared by crews today are excessive in the case of those fortunate in making large catches and insufficient for those who render the same service but have a run of bad luck. . . . It is in the interest of the industry and of the public that gambler's chance should not play so large a part in the fishermen's wages." The efficiency of steam trawling and its modern wage structure, when properly regulated, could provide a means of going forward. "A much more healthy condition now prevails on the case of the trawlers, in which members of the crew receive wages at the rate of $40 per month for the trip, and a small share (7 percent) in the profits. If this method, or a similar plan, could be extended to the schooners and other vessels in the fishing fleet, it would stabilize the industry without removing entirely the chance to share in the luck of the ship."[57]

As for the wholesalers and retail dealers, the committee came to similar conclusions: "Viewed from the standpoint of the general public, the time has come when the consumer cannot be expected to support the old-fashioned retail fish market which will not long survive the closer competition of modern business conditions." While the behavior of the cartel and its confederates broke the public trust and inflicted unjustifiable economic costs on fishermen and consumers alike, new business models could reform an antiquated industry to prevent inefficiencies that kept much-needed food from Boston residents.[58]

Whereas Boston's dealers faced withering scrutiny, Gloucester's salt fishery, consolidated under the Gorton-Pew Fisheries Company, avoided further investigation although Gorton-Pew had also subsumed all of its competitors. Expanding its holdings of fishing and other ancillary firms after 1908, they enjoyed almost exclusive power to set salt fish prices in New England markets. As a result, the recess committee found that "present prices are undoubtedly higher than are warranted by the conditions in the industry during this past year." Despite finding undue market influence, however, the committee deemed it "inadvisable to recommend that any action be taken by the General Court at this time."[59] All that was needed to spur competition, the committee maintained, was reduction in the protective federal tariff levied against foreign salt fish products, a policy change beyond the authority of the Massachusetts General Court.

Change came later as the various criminal cases wended their way over the next five years through Massachusetts courts. The commonwealth initially agreed that in exchange for opening the Boston Fish Pier to wholesalers and retailers, the BFMC would not be broken up. As a result of charges leveled under the 1914 Clayton Antitrust Act, however, which mandated the breakup of monopolies, cartels, and trusts, federal authorities dissolved the firm.[60] As for the individuals involved, one judge indicted several of BFMC's key operatives, imposing jail sentences that sparked a slew of appeals that dragged on into the mid-1920s. By then, Boston's fish consumers were eager to see the prison sentences upheld. On April 7, 1923, the *Boston Globe* ran the provocative headline, in large font, "Fish Trust Men Have Best Island Affords: Fourteen Begin [Jail] Terms, Segregated in Modern Prison, Formerly Used for Women Prisoners ... Drug Addicts Only There ... First Meal Served to Them Is Fried Fish." Spanning two pages, the story recounted how the convicted financiers and fish dealers walked to the boat that would carry them to Deer Island Prison: "One of those sentenced to five months and to pay a fine of $500, tried hard to set a good example of cheerfulness but, though his banter brought forth smiles, for the most part, the men were doing some hard thinking. They had been sentenced to hard labor and more than one was curious as to what type of labor he would be called upon to perform."[61] After four years of court trials revealing to all how these individuals manipulated wartime prices and preyed on consumers, the *Globe*'s headlines likely offered the public some satisfaction.

The Moral Uncertainty of Modern Fishing

By the time the "Fish Trust Men" reported to their prison terms on Deer Island, New Englanders had found good reason to distrust Boston's modern, mechanized fishing business. Steam trawlers and industrial processing brought fish to market far more efficiently than the more venerable schooner-based hand line fishery, but the capitalists controlling Boston's modern fisheries ruthlessly pursued profits at public expense like other so-called Robber Barons. Such distrust did not extend to Gloucester's salt-fish fishery, however. Even though it had experienced just as much consolidation, market manipulation, and abuse of commercial power, Gloucester's market was still reviewed favorably by regional consumers. State investigators gave Gloucester outfitters a pass during their investigations and subsequent reform initiatives, perhaps due to Gloucester's political influence in the state and its industry's traditional appearance. Indeed, the differences between the two fisheries were chiefly the catching and processing of the fish. For most New Englanders, however, Boston's modern, mechanized, industrial fishing firms proved too novel, too efficient, and too powerful to remain unchecked. The 1919 investigations and subsequent indictments not only curbed BFMC's market power, swelled by the defeat of the Gardner bill; they impugned the trustworthiness of trawl fishing and helped to revive the ideal of New England fishing entrenched in the books of Kipling and Connolly—a spirit bowed but unbroken.

CHAPTER 3

MASKING INDUSTRIAL REALITIES

Schooner Racing, Industrial Fishing, and Federal Aid, 1919–1939

As Boston's fishing cartel faced indictment for corruption, community leaders in Gloucester learned of a new opportunity to celebrate New England's traditional fishing culture and heritage on an international stage. Early in 1920, the editor of the *Halifax Herald,* William H. Dennis, challenged Nova Scotia's fishing rivals to a series of international schooner races. Crews from Canada and the United States would compete for a prize cup and five thousand dollars. Unlike the America's Cup races, fishermen would race according to fishermen's rules, with only vessels actively fishing allowed to participate. Yachtsmen and yachting rules would have no place in this competition.[1]

It was audacious challenge based on a rivalry that American and Canadian fishing captains had held onto for generations. Once fully loaded on the banks, captains raced their vessels home, for whichever vessel arrived in port first was likely to fetch higher prices for its catch. In the nineteenth century, the schooner races caught the attention of vessel designers, such as Tom MacManus, as a means to assess refinements in design. For the most part, though, these races had no rules, no conventions, and no pretense to be anything but a mad-dash for the best price. As historian Wayne Santos describes it, it was a working-class sport that had caught the attention of yachtsmen seeking authenticity in their avocation.[2] Dennis may have thought that his challenge would be left hanging. Many Nova Scotia fishermen had fished out of Gloucester or knew someone else who had done so. They were also aware that schooner fishing in the United States was fading. Given the qualification criteria, few if any American vessels would be able to race.

Indeed, American indifference almost prevented the race from taking place. After waiting weeks for an American captain to step forward, it appeared that Gloucester's race committee would have to decline Dennis's challenge. According to local lore, however, the schooner *Esperanto* arrived in Gloucester after ten weeks with a hold full of fish from the Scotian Shelf fishing banks lying off Nova Scotia's eastern shoreline. The deadline to accept the challenge had dwindled to a few hours. Gloucester's race committee met the vessel at the dock, and the captain and crew agreed to compete. Within a day or two, *Esperanto* had been offloaded, cleaned up, and set sail for Halifax to take up the gauntlet. *Esperanto* won the first international schooner race between the United States and Canada handily, taking the first two races in a best-of-three series. Yachting magazines and fishing trade journals picked up the news of *Esperanto*'s win and spread it throughout the U.S. East Coast. The win sparked fervor among Canadian and American boosters who launched the schooner races for which Gloucester is now famous.[3]

Importantly, *Esperanto*'s victory contrasted starkly with the news out of Boston. Almost daily, newspaper reports detailed the ongoing investigations into Boston's frozen fish industry, highlighting its corruption, monopolistic business practices, and wartime profiteering. Coverage had become standard fare for the *Boston Globe*. For those who lionized Gloucester's vanishing schooner fleet, like novelist James Brendan Connolly—who had crewed aboard the victorious *Esperanto*—the win clarified the differences between New England's burgeoning but soulless mechanized fisheries and its declining but stalwart and heroic schooner fisheries. For casual readers, *Esperanto* and her crew kept New England's traditions alive for a little longer in the face of international and internal competition.

A prominent story published in *National Geographic* magazine in 1921 further reinforced the traditional image that set New England fisher men apart from monopoly, mechanization, and corruption. Perhaps in response to four years of fishing scandals surrounding the mechanized fishery, Frederick Wallace penned "Life on the Grand Banks" in what can only be seen as an attempt to restore the North American fisherman to the pantheon of iconic Americans. Resurrecting earlier tropes from Winfield Thompson, Edwin Gould, and others, Wallace invoked generalizations, objectifications, and racialized imagery to present fishermen as timeless, racially and ethnically pure Americans standing apart from the industrializing world. "The deep sea fishermen are a distinct type peculiar to the North American coast," he wrote. "Racially they are from the sturdy pioneer breeds of Highland Scotch, Hanoverian German, West Country English, and West Irish which settled in Newfoundland, eastern Canada, Maine, and Massachusetts." Echoing Thompson, Wallace presented the fisheries in a distinctly celebratory, antimodern frame: "Nowadays, the men who built and sailed the American sailing marine . . . [have] disappeared." Those who Wallace identified as newcomers largely appeared cut from the same cloth. "Their places have been filled by those of their breed who have succeeded in resisting the allurements of the shore industries and the cities." While he acknowledged that Portuguese, Scandinavian, and "native-born Americans" had entered the fishery, to Wallace, all shared the healthy physique of rural people raised far away from the ills of modern cities. "Physically, your American deep-sea fishermen are strong-muscled and able to endure hardship. They are not slum or city products, but are mainly raised in sea-coast villages of the Canadian provinces and Newfoundland. . . . Clean air, good, wholesome, food, and hard work create a sturdy, hard-muscled youth who usually breaks away to sea in a Bank fishing vessel ere town lads are through grammar school."[4]

For Wallace, dory fishing, a hazardous job under the best of conditions even in the 1920s, honed fishermen into icons: "It is this dory fishing which makes the American fisherman, and by that term I include Canadian and Newfoundlander, a distinct type from his colleagues in other countries, and adds to his vocation a hazard and labors which calls for certain sterling qualities to surmount." These qualities helped fortify Wallace's idealized fishermen against the evils of modern life: "It is a peculiar fact that the North American fishermen, of all white fishermen, has stood out longest against modern innovation in fishing methods and equipment." In part, this was due to the symbolic role of the fishing vessel in reaffirming political and social values that many saw as under threat. "Every Bank fishing schooner is a sort of seafaring democracy," he continued. "The crew works the ship on a cooperative basis, with the skipper as sailing and fishing boss." Ultimately, these attributes produced an image of the fisherman as the perfect citizen, serenely capable of withstanding a changing North American society. "The average deep-sea fisherman of today is merely a healthy, level-headed, intelligent class of skilled worker . . . [with] daring, initiative, skill in seamanship, and ability to endure long hours of heavy labor and the rigors of seafaring in small vessels during the varying conditions of weather on the North Atlantic."[5]

Grudgingly and with a dirge-like air, Wallace acknowledged the transformations overtaking New England's fisheries. Modern life threatened not only the fishermen he celebrated but also American consumers and citizens. More importantly, though, at the end stood his mournful admission that change seemed inevitable: "The age of the clipper ship and the seamen who sailed them is gone, but in the American Banksman we find the smartest sailing craft and the smartest sailormen afloat today. But the steam and motor trawler is coming into the American fisheries and many of the tall-sparred schooners are having their sails and masts cut down and internal combustion engines installed."[6] Regardless of what New England's fishermen actually felt about their industry, and despite the labor unrest, corruption trials, and investment in new technology, Wallace's article illustrates how public perception of how New England fishermen looked, behaved, thought, felt, and fished had changed little over the previous half-century. Despite all evidence to contrary in an industry that was anything but static, Americans at large refused to see how fishermen themselves were adapting to a changing world.

Expanding Impacts of Trawl Fisheries

If *Esperanto*'s win and arguments like Wallace's encouraged schooner boosters, everyone admitted that time was running out on New England's and especially on Gloucester's schooner fleet. Since 1908, when the U.S. Bureau of Fisheries (USBF) began recording data specifically from the otter trawling fleet, the fleet's role in the region's fisheries had escalated dramatically. Bay State Fishing Company, following the success of *Spray* in 1905, had fitted out five more steam trawlers by 1912. Another steam trawler, owned in New York and launched around the same time, landed its catch in Boston. Three more trawlers came into service in 1913, and in 1915 a fleet of twelve vessels operated out of Boston and, to a lesser extent, Gloucester and Portland. Steam trawling expanded during World War I, despite the navy's requisition of some of the trawlers for wartime service. By 1919, Boston's otter trawling fleet totaled twenty-five American and two Canadian vessels, which offloaded catch in New England ports per temporary wartime agreements.[7]

New England's trawler fleet itself grew rapidly, but its share of haddock and cod catch grew astronomically. The number of trips the vessels took and their percentage of the total landed catch reveal the expanding impact of steam trawling. In 1921, trawlers took 346 trips to land 40 percent of New England's total haddock catch. In 1922, steam trawler activity rose sharply: 578 trips landed 35.9 million pounds (16,288 mt) of haddock and 11.2 million pounds (5,081 mt) of cod. By 1923, otter trawlers had come to rival the landings of the venerable schooner fleet. In that year, 6,535 trips made by schooners landed 58.2 million pounds (26,406 mt) of cod and 73.7 million pounds (33,439 mt) of haddock. In contrast, 33 large otter trawlers made 665 trips landing 14.9 million pounds (6,760 mt) of cod and 35.5 million pounds (16,107 mt) of haddock. The catch of those trawlers constituted just over 48 percent of the total haddock landed and 42 percent of its total value. In one-tenth the number of trips, steam trawlers pulled in roughly half the landings of the schooner fleet.[8] With the vast majority of steam trawlers operating out of Boston, observers could not ignore how mechanization was changing the nature and heart of northwest Atlantic fisheries.

Otter trawling proved so successful in fact that it almost killed itself off. While the Commonwealth of Massachusetts pressed charges against corrupt dealers and processing firms, the fishing industry fell into a tailspin of its own making. In 1920, a glut of fish on the market crashed prices and eroded the share values of firms that first engaged in steam trawling. According to the business journal *United States Investor*, "The figures paid for [shares of fishing companies in Boston] . . . must have brought a shock not only to unfortunate investors . . . but also to the New England mind which has always regarded fishing as one of the mainstays of New England industry."[9] Sales plummeted as fish caught by brand-new steam trawlers flooded the market and as American consumers, freed from wartime restrictions, returned to eating beef. In turn, stocks in fishing firms declined across the board, compelling the *United States Investor* to report, "Those companies which depend almost entirely upon the earning of steam trawlers for income seem to have been hit . . . for the reason that the market for fresh fish is overstocked and the operating costs of steam trawlers have become too high to permit steady profits."[10] Bay State shares fell, as did shares of East Coast Fisheries Company, which had purchased twenty trawlers from the French government in 1918, just in time to watch its share prices crater to four cents on the dollar. The title of a subsequent multipage *United States Investor* story made the point quite clear: "The Crash in the East Coast Fisheries: New Fangled Notions of Fishing Come to Grief."[11] Too many trawlers, too little demand, and too high a level of capitalization undermined New England's first foray into mechanized fishing.

In contrast, Gorton-Pew survived the crash, and not just because of its well-established, international, and diversified markets. Equally important to the *United States Investor*, "this company is in the hands of real fishermen. It was not promoter-managed." Bay State Fishing Company also managed to survive both the crash and the indictments, but just barely. With share values dropping to roughly ten cents on the dollar, by 1920 the firm was forced to undertake extreme economizing measures, including laying up ten of its fourteen trawlers to cut overhead costs. Built at a cost of $200,000 each, the value of a trawler at the time of the crash according to one estimate was merely ten cents on the dollar, assuming a buyer could be found.[12]

Resolution of Boston's fisheries scandal did little to help Bay State. Its forced breakup as a result of the firm's involvement with BFMC during the war, along with the subsequent court rulings designed to deny any one firm market dominance, further undermined viability. Additionally, some felt that Bay State had spent too much money on efficiently catching large volumes of fish. Wage structures, headlong investments in machinery, abandonment of the traditional share system in favor of a modern industrial financial model all led trawling firms into trouble. Again, according to reporting in the *United States Investor*, "There is an absence of the partnership principle as in sailing vessels, so that if the prices, or the catches, are low, the owner of the steam trawler assumes more of the burden than does the owner of the sailing vessel."[13] The lesson was clear: by adopting financial models and principles borrowed from manufacturing, New England's first-generation trawling firms suffered brutal galvanization in the market crucible of 1920.

Revolutionizing an Industry, Saving a Young Fishery, Exploiting a New Species

Unlike traditional operations where few outside the fisheries understood business well enough to gamble funds, the modern industrial foundations of steam trawling made it easier for troubled firms to find buyers and investors. With the 1920 crash, Bay State faced bank debts amounting to $600,000 and unpaid war profit taxes totaling $500,000. Sensing a bargain, that same year Boston financier and corporate lawyer B. Devereux Barker took over the firm. He renegotiated the tax bill down to $130,000, paid down the bank loans, and adopted new marketing styles to expand sales. Rather than sell fish whole—that is, gutted with head and tail remaining—Barker sold just the edible portion of the fish in cuts called fillets. He raised prices to cover the added costs of fillet processing but found consumers willing to pay the increase in exchange for the convenience.[14]

Other trawling firms found similar salvation. In 1922, a financier who specialized in corporate resurrections, Ira Cobe, assumed control of the bankrupt Atlantic Coast Fisheries. He brought in Harden Taylor, chief technologist at the USBF, to develop quick-freezing methods for fish. Embracing the "Taylor Method," which used brine and contact plates to freeze fish quickly, Cobe invested $1 million in a processing plant in Groton, Connecticut. Halfway between Boston and New York, the plant provided easy access for Boston-based vessels to land catches for processing and sale in both markets. Cobe appears to have succeeded for by the late 1920s, Atlantic Coast Fisheries operated the second largest trawler fleet behind Bay State's.[15]

Even Gloucester benefited from the application of modern business financing to fishing. While Barker marketed fish fillets, and Cobe and Taylor developed quick-frozen fish for the New York market, Clarence Birdseye commercialized his freezing process for fish production. In his first attempt in 1922, Birdseye used chilled air to freeze fish, but the product failed to garner consumer interest, and his firm went bankrupt. Two years later, he developed a new method that froze packaged fish whole using pressure as well as cold air. This method proved far more successful. Birdseye formed the General Seafoods Corporation to process haddock landed in Boston for sale in supermarkets. In 1925, he moved his processing plant to Gloucester and in 1929 sold it to the Postum Corporation, soon to be renamed the General Foods Corporation, for $22 million.[16]

A profitable revolution in processing, freezing, and distribution could only work on an industrial scale, however. That required high volumes of fish moving steadily across Boston Fish Pier and into processing plants. For the new mechanical processing methods, haddock proved ideal. The fish's firm, tasty, clean-smelling white flesh was sufficiently oily to withstand freezing better than leaner species like cod. Furthermore, individual adult fish grew within a relatively consistent size range, which facilitated designing mechanical processing machines. Most importantly, haddock was abundant. While schooners had long targeted inshore haddock stocks for the fresh fish market (landing about 60 million pounds [27,200 mt] a year between 1880 and 1918), the salt fishery avoided haddock because its oilier flesh did not lend itself well to salt curing. As a result, offshore stocks, even those found ninety miles east of Boston on Georges Banks and in the Great South Channel separating Georges from Cape Cod, had experienced far less fishing pressure than cod stocks had. By 1920, as steam trawling began to expand in earnest, haddock resources stood healthy, abundant, and strong—as far as anyone could guess. For trawlers, haddock was an obvious choice. Boston's haddock fleet boasted annual landings of well over 80 million pounds (36,298 mt) by 1920, 107 million pounds (48,548 mt) for the year in 1924, and 174 million pounds (78,947 mt) for 1927.[17]

In part, the dramatic increase in haddock landings stemmed from changes in net design. Originally, steam trawlers attached otter doors to the net's wing-ends, or the widest parts of the net itself. In 1923, however, Raymond C. Mudge, a designer working for Bay State, filed a U.S. patent application for his adaptation of a European Vigneron-Dahl trawl. This gear greatly expanded the area swept by dragged nets by lengthening the otter trawl's bridle cables and by positioning the otter doors between the bridle and the net mouth. According to the designer, "By the virtue of the long connections between the mouth of the bag and the otter boards, fish are collected from a much wider area, namely, an area bounded on the sides by the otter boards since the boards moving through the water frighten the fish and cause them to deflect their course toward the mouth of the net." Additionally, Mudge added a "tickler" chain along the ground cable of the net that "bows rearwardly somewhat in advance of the bottom . . . beneath the overhanging top of the bag, [and] drags on the ground and directs the fish upwardly into the mouth of the bag."[18] These two adaptations greatly expanded the efficiency of the dragged nets: fish scared by the door's ground plume or kicked up by the tickler chain swam toward the center of the net's mouth and were taken.

Large otter trawlers quickly adopted the Vigneron-Dahl trawl (Mudge's patent notwithstanding) and started landing a lot more haddock. In 1925, the USBF reported that the otter trawl fleet landed over 44 million pounds (19,963 mt) of haddock. Most of those fish came from the Great South Channel, as its sandy bottom lying between Georges Bank and Nantucket was ideal for deploying otter and Vigneron-Dahl trawls. Two years later, twenty-six vessels landed almost 70 million pounds (about 31,760 mt) of haddock, mostly from the same area.[19]

By 1928, it was clear that Boston's haddock fleet had grown dramatically. In light of such growth, the USBF took a fresh look at the otter trawl fleet. Directed by new assistant in charge, Reginald Hobson Fiedler, investigators found that not only were Boston's long-distance steam trawlers using both otter and Vigneron-Dahl trawls but that Boston's haddock trawl fishery had come to include forty-three vessels, registering ninety-one tons or more burthen, using Vigneron-Dahl and traditional otter trawls exclusively. Additionally, long-distance trawlers now sailed from Gloucester and Portland, Maine. In total, Fiedler's investigations found that these vessels clocked over 1,000 fishing trips, totaling more than 7,700 days, landing roughly 87.5 million pounds (39,700 mt) of fish. Eighty-seven percent of that total (roughly 75 million pounds [34,029 mt]) was haddock. Less than a decade after the New England Fish Exchange indictments and the fish market implosion, Boston's heavy trawler haddock fishery had recovered, and thrived.[20]

Others also thrived as trawls reshaped New England's fisheries. In addition to the well-known Boston fleet, eighty-nine vessels between twenty-one and ninety tons, referred to as "draggers," took 10 percent of the region's total fish catch in 1928. These "medium" vessels logged over 1,000 trips, totaling roughly 6,100 days absent, and landed about 30 million pounds (13,611 mt) of fish, of which roughly 25 million pounds (11,343 mt) were haddock. Even smaller day-boats operated with dragged nets. One hundred and twenty vessels between five and twenty tons burthen took 1,181 trips totaling 4,092 days and landed roughly 11.3 million pounds (5,127 mt) of fish. These smaller boats—referred to as "flounder draggers"—constituted the relatively new flounder fishery that had been identified in the late 1880s but only successfully pursued after World War I. In addition to vessels using otter trawls proper, Fiedler's people also found a large number of vessels turning to the more efficient Vigneron-Dahl trawl. Six vessels over ninety tons and thirty vessels between twenty and ninety tons used the new gear. This fleet of thirty-six boats landed roughly 6 percent of the New England's total fish landings in 1928, taking 354 trips, comprising around 2,200 days, again targeting mostly haddock.[21]

The development of Vigneron-Dahl trawls, combined with peculiarities coming out of the 1919 indictments, created unique opportunities for New England's smaller vessel fisheries. By using sweeps and doors to kick up a sand cloud, Mudge expanded the gear's catch rate without adding much to the gear's weight and drag through the water. Consequently, gasoline-powered vessels, which slowly entered the fishery beginning in 1900, could now tow nets with great effectiveness. Additionally, the courts gave smaller New England vessels a unique niche in the market to sell their expanding catches. The commonwealth's provision that all fish landed at Boston Fish Pier must be sold in open auction—that buyer's must buy fish off their own vessels—not only removed Bay State's power over the market; it also removed the potential influence of larger vessels that could catch a lot more fish.[22] Smaller-scale operations now obtained the same prices, even at smaller volumes, that the larger fishing firms had once exclusively enjoyed. The legal settlement leveled the playing field by guaranteeing New England's small fishing vessels and their larger competitors the same market position. Thus, the settlement created a unique fishery in American waters. With more efficient gear, open access to markets, and a level rules, New England's small fishing operations could, and would for the rest of the century, effectively withstand the forces of combination and consolidation that shaped fisheries in other parts of the country. In doing so, they would come to enjoy over time a cultural resonance far exceeding their economic influence. As the next few decades would reveal, the small fisheries would also outlive the larger operations that had pressured them before 1919.

Academic Legitimation of New England Fishing Icons

Frederick Wallace was not alone in mourning the changes in New England's beloved industry. While he and his peers wrote pieces for the popular press, prominent academics similarly embraced the stock image of the New England fisherman and mariner. Certainly, George Brown Goode's monumental survey for the U.S. Fish Commission in the 1870s and published through the 1880s had provided an awkward, quasi-anthropological, quasi-natural history foundation for analyzing New England's fishermen.[23] Successors trained in the social sciences and humanities, however, lent popular narratives academic credentials by openly embracing romantic representations. Perhaps more than any others, scholars such as Middlebury College's Raymond McFarland and Harvard historian Samuel Eliot Morison provided academic legitimacy to the popular imagery of New England fisheries and fishermen. By the 1920s, the romanticization of New England's maritime past shaped scholarly and even scientific understandings to a remarkable extent. Those understandings would play a critical role in regional fisheries management when people openly asked if the modern haddock fishery was hitting stocks too hard.

Raymond McFarland's 1911 historical study of New England's fisheries owed much to late nineteenth-century Boston Brahmin and Harvard historian Francis Parkman. Through a series of books focused on the

French and Indian War, Parkman presented the European struggles for North American empires as a conflict that pitted Protestant English colonists against Catholic French settlers and "savage" Native Americans. While acknowledging the skills and prowess of notable individuals in each episode, Parkman believed that the common people who worked and fought in the North American wilderness played as critical a role in the bloody conflict as the elites who allegedly led them. In grand, sweeping prose, Parkman infused his colonial histories with a romanticism based on the honor of the common man that belied his own elite status in Boston's high society. Indeed, "doing" was such an important aspect of his historical perspective that despite chronic health problems that laid him up for weeks, Parkman traveled to colonial battlefields and historic sites which in the 1870s required rigorous treks into wilderness similar to those made by his protagonists.[24]

McFarland rooted his study, *A History of the New England Fisheries*, squarely on Parkman's heroic vision of New England's common man—this time, its fishermen. In contrast to George Brown Goode's monumental five-section, seven-volume *Fish and Fisheries Industries of the United States* (1884–87), McFarland sought to address the poor understanding of the role New England fisheries played in the nation's heritage and founding. Stating frankly that, "no literature ... adequately sets forth its history and value," McFarland sought to include fishermen as subjects for veneration within contemporary regional celebrations of the colonial past. But like Gould and Charles Francis Chamberlayne before him, McFarland's *History* also looked to inform public policy. In particular, McFarland wanted to remind U.S. representatives working to resolve fisheries disputes among the United States, Canada, and Great Britain then underway at The Hague of the social and cultural value of the fisheries. Grand in its view, appearing conveniently comprehensive, McFarland's work was less a history of New England fisheries than a determined exposition of the "importance of the New England fisheries from pre-colonial days to the present." Claiming that "the Story of the Fisherman" (invoked not as individuals but as a type) had yet to be written, McFarland believed that "it could be made a volume of keen interest, a narrative of heroic daring."[25]

Academic reviewers were not impressed with the book's scholarly foundations, however. Historian Carl Russell Fish took McFarland to task for the work's polemical and antiquated nature: "On the political side, the book is weak. The author is in favor of a revival of the bounty system . . . and of a continuance of protection. . . . It rather seems that he is naively unconscious that any difference of opinion has existed within the United States on the subject of [fisheries] protection. . . . In light of the wide grasp of the other problems related to the fisheries, this gives an incomplete and devitalized effect."[26] James Scott Brown, a lawyer involved with the international negotiations, proved less restrained in his criticism. He flatly rejected McFarland's *a priori* assumption of the current importance of New England's fisheries: "The New England fisheries are not so important at the present day as they were from the Colonial period to the end of the Civil War. Industry and commerce have developed to an unexpected degree and have over-shadowed the value and role of the fisheries. They, nevertheless, continue to furnish employment to a large section of the community, but they are not, as formerly, the chief characteristic of New England, and the chief source of livelihood of its people."[27] McFarland's study provided romantic, stirring, and inspiring reading but was inadequate as academic scholarship. These reviewers agreed that *A History of the New England Fisheries* was suitable only for general audiences.

All that changed, however, following the Gardner bill hearings, the birth of the heavy trawler fleet, and the indictments at Boston Fish Pier. In 1921, Samuel Eliot Morison published his *Maritime History of Massachusetts*, which painted the Bay State in the same grand narrative strokes used by McFarland and Parkman. This time, however, Morison's heroic story of Massachusetts's mariners and fishermen achieved scholarly as well as popular acclaim and robed the romantic image of the New England fisherman in academic legitimacy. As future federal fisheries hearings showed, Morison's study would frame how generations of politicians viewed the fisheries, and in doing so, shape fisheries regulations well past midcentury.[28]

Like Parkman, Morison venerated those who "did" more than those who "thought," with a sweeping writing style and grandeur lending his exaltation of humble New England fishermen a romanticism more commonly found in popular representations. For Morison, New England fishermen personified, in act and appearance, the region's solid tradition, heritage, and conservative values. Like many popular writers of the time, Morison proclaimed, "Fishermen had their own customs and costumes, types and traditions which were handed down from generation to generation." A fisherman's son was "predestined to the sea," for "as soon as he could walk, he swarmed over every Banker or Chebacco boat that came into port, began hand-lining [for inshore fish such as] cunners off wharves and ledges, and begging older boys to teach him to row." Fishermen harbored only the most modest of ambitions in Morison's world as well: "To save enough to acquire a fishing vessel, and live ashore off her earnings, was his highest ambition." They faced an old age with equal contentedness, for "when rheumatic arms could no longer haul a sheet or cable, and eyes grew dim from straining through the night, fog, and easterlies, [the fisherman] retired from deep waters, and puttered about with lobstering, shore fishing, or clam digging."[29]

Docile, static, traditional, Morison's fishermen could not also be innovators. In vessel design, for example, Morison claimed that "fishermen, the most conservative of seafarers, seem to have made no improvement in their models until after 1815. Methods were unchanged . . . [and] the only innovation of the Federalist period was [fishing in] a wider [geographic] range." Despite writing the book at the same time that Bay State Fishing Company members faced indictment, Boston halibut fishermen struck at the Boston Fish Pier, and other fishermen threatened to strike over the right to unionize, romanticism allowed Morison to conclude that Massachusetts "never had a native deep-sea proletariat."[30] Instead, his fishermen, like McFarland's, stood as bastions against demographic change, innovation, and novelty despite the dynamism of the contemporary industry.

McFarland's and Morison's works embedded the popular myth of New England fishing, now utterly divorced from industrial reality, firmly in the academic canon. Indeed, these two works, but particularly Morison's, became standard texts for readers interested in New England fisheries for the rest of the century. Their interpretations would appear repeatedly in the press and in governmental hearings beyond the 1970s. Lobbyists and regulators relied on these two convenient, engaging, and accessible sources of information, which colored for the rest of the century New England's fisheries regulatory discussions with a romanticism far greater than popular literature, art, or film had ever done before.

Nostalgia, Racing, and Industrial Redemption

After several years during which race organizer wrangling stifled schooner racing, the Northeast again became caught up in racing mania in the late 1920s. Behind the revival stood Ben Pine, a Gloucester businessman, Newfoundlander by birth, who emigrated to the United States in 1893, fished for a season or two, then turned his hand to vessel ownership and the junk business. Crafty, smart, and able to make good deals for himself, Pine found commercial success in the 1910s with the twelve-vessel swordfish and mackerel fleet he co-owned with Joseph Langford. By 1920, however, Pine had caught the schooner racing bug. He commissioned the construction of the aptly named schooner *Mayflower* specifically to race (and fish only enough to qualify). The following year, Pine borrowed and captained with some skill the vessel *Philip P. Manta*. In 1922, he had another schooner designed and built primarily for racing. Bearing the historically and ethnically charged name *Puritan*, the schooner foundered shortly after launching. Pine's efforts continued for a decade, despite setbacks, as throughout the 1920s, Pine sought to rekindle Gloucester's passion for schooner racing as a sport and tourist attraction. As Gloucester's schooner fleet grew increasingly dormant, and as more draggers landed fish at Gloucester's piers, Pine saw the races as a chance to put his town on the map and to keep its idealized past alive.[31]

Pine's dogged persistence finally paid off in 1929. Although the Canadians still refused to race due to legal squabbles, Gloucester's tercentenary provided a nostalgic milieu ripe for another series of schooner races. To support the events, Pine solicited donations, ironically from the businessmen whose modern fleets were killing off the schooners. These benevolent citizens and their industrial firms, eager to boost the tourist industry, shelled out ample funds to provide unfamiliar bystanders with a crash course in New England's fishing heritage. Boston Fish Pier's O'Hara Brothers entered the race, sponsoring the schooner *Shamrock*. Kellog Birdseye, brother to Clarence, provided cash funding. So too did the Postum Company (which had purchased Birdseye's operation in 1929) and its subsidiary, the General Seafoods Corporation. For these firms, the 1929 races provided a glorious opportunity to don the mantle of tradition and perhaps rehabilitate their industry's tarnished image.[32]

The irony of Gloucester's schooner race and those who funded it was not lost, however. An anonymous piece in the populist yachting magazine *The Rudder* pointed out that if Gloucester intended to celebrate its fisheries, a fleet of draggers should race, "pulling along with Diesel engines wide open and perhaps the forestaysail set if the wind slant was favorable."[33] Still, Gloucester's anniversary provided a perfect moment when past and present, myth and reality, could blur, and people could bask in the warm light of nostalgia.

Hypocritical or not, the races proved to be a success. In the following year, Pine set out again to build a schooner that could defeat the Canadians and their prized schooner, *Bluenose.* In the fall of 1929, Pine and local businessman Joe Mellow founded the firm Gertrude L. Thebaud, Inc., to build and race a vessel designed to defeat *Bluenose* in the next international schooner racing series. Pine and Mellow again solicited donations from the same industrialists marginalizing Gloucester's schooner fleet. Wetmore Hodges, an executive working for General Seafoods, contributed five thousand dollars. Retired New York insurance executive Louis Thebaud—after being regaled by retired fishing captains in Gloucester's Master Mariner's Hall—ponied up thirty thousand. His wife, Gertrude, contributed ten thousand dollars, and two of his relatives each donated another ten thousand. Launched in March 1930 to great fanfare, the schooner *Thebaud* lost handily to *Bluenose* in two straight races in 1931. The outcome hardly mattered, however; for Pine and Gloucester's supporters, the schooner races had kept the schooner fishery alive, put Gloucester and its heritage on tourists' maps, and even added a couple of new vessels to the fleet.[34]

Confronting the Consequences of New England's Modern Haddock Fishery

While Pine regaled the nation with the spectacle of schooner racing, others wondered how long the heavy trawler fleet could continue to hammer haddock stocks before they gave out. Since the 1920s, the chief of the USBF's New England haddock investigations, Henry Bryant Bigelow, had been aware of European discussions about overfishing in the North Sea. Bigelow had been the U.S. representative to the International

Coalition for the Exploration of the Seas (ICES), the international body of European fisheries scientists then concerned with managing Europe's industrial fisheries. Following World War I, work by British researchers Michael Graham and E. S. Russell had provided statistical evidence supporting the idea that modern fishing could deplete stocks of marine fish. While the most prominent Canadian fisheries biologist, Bigelow's close friend A. G. Huntsman rejected Russell's and Graham's theory, Bigelow began to collect fishing data measuring the effort fishermen expended in catching their fish, also known as effort data, in 1924. That data, in turn, would allow New Englanders to run analyses similar to those run in the United Kingdom.[35] By the late 1920s, as Graham's and Russell's studies began appearing in international journals, Bigelow had established the data foundations to explore whether New England haddock fisheries were unsustainable. Still, Bigelow appeared torn between the conclusions presented by Russell and Graham, and Huntsman's rejection. Despite ordering the collection of effort data, Bigelow never openly supported the notion that human fishing could affect stocks of marine fish. Furthermore, even as late as 1930, while acknowledging the help his new field of oceanography could provide in mitigating fisheries fluctuations, Bigelow never actually mentioned the term "overfishing" in his famous article, "A Developing Viewpoint of Oceanography," nor did he do so in his book, *Oceanography*.[36]

In 1930, Bigelow stepped down from the haddock investigations to launch the Woods Hole Oceanographic Institution. The research foundation he had laid, however, quickly proved invaluable to his successor, William C. Herrington, a scientist newly arrived from the West Coast. Unlike Bigelow, a Boston Brahmin steeped from birth in the ideology of New England fisheries, Herrington was a Midwesterner standing outside New England's cultural milieu. Perhaps this was why he could conceive of situations where modern fishing practices might deplete marine fish stocks. Cultural inclinations aside, however, Herrington's previous work with William F. Thompson on the International Pacific Halibut Commission produced results supporting Russell's and Graham's conclusions.[37] More importantly, Herrington's work with Thompson illustrated the powerful role that rigorous fisheries science could play in effectively managing and sustaining a fishery. With these experiences under his belt, he took up his position in Woods Hole, Massachusetts, and set about addressing whether New England's otter trawlers were overfishing local haddock stocks.

Within the year, Herrington saw that the fishery would soon be in trouble. In his 1932 article calling attention to the destructiveness of the heavy trawler haddock fishery, Herrington pointed out that New England's fishing industry was not what people thought it was. Herrington observed that between 1905 and 1930, New England fishermen shifted from "gallant sailing schooners," using baited hooks and lines, to diesel-powered vessels and steam trawlers using otter and Vigneron-Dahl trawls. Within those twenty-five years, the schooner fleet dwindled to only 115 schooners, while the number of trawlers grew to a remarkable 323 vessels of all sizes. More impressively, the groundfish fleet expanded its annual landings by more than 320 million pounds (145,191 mt) over the same period of time.[38]

Herrington had few problems with a profitable and successful fishery. Indeed, the Bureau of Commercial Fisheries (BCF), the successor federal agency to the U.S. Bureau of Fisheries, supported just such an outcome. Furthermore, he dismissed complaints that otter trawls destroyed ocean floor habitats, claiming instead that "the uninjured [ocean-floor dwelling, or benthic, organisms taken up in otter trawls] . . . resume their life as before and any crushed scallops mussels, etc., [will] be eaten by fishes or by animals on which the fishes feed."[39] What Herrington found appalling was the amount of waste that went into the mechanized catching of groundfish. For the roughly 37 million individual haddock New England vessels landed in 1930—that made up the 150 million pounds (68,058 mt) caught—trawlers threw overboard dead an additional 70 to 90 million undersized, juvenile haddock too small or too mangled for market. Assuming these juvenile fish weighed half the average size of the fish landed, Herrington's figures point to discards in the 160 million pound (72,595 mt) range—10 million pounds (4,537 mt) more than the total landed.[40]

With discards amounting to as much or more than the landed catch, Herrington felt compelled to find a better way to fish. He noted, "Considering that the destructiveness of the otter trawl fishery has been evident since its first introduction into the fishery, it may seem strange that no attempt has been made to modify the gear in order to overcome this characteristic." Herrington offered two possible explanations for the persistence of such wastefulness. The first stemmed from the fact that New Englanders used small-meshed nets designed for European fisheries targeting smaller fish. The second rested on cultural perceptions of New England fishermen: "Once having become accustomed to a certain type of trawl, the fisherman's conservation nature [i.e., unwillingness to change] has prevented him from attempting modification to permit the escape of small fish." Imprisoned in his own traditions, the New England fisherman "has come to consider that the taking of large numbers of small fish is an essential characteristic of the otter trawl."[41] For Herrington, the twin evils of tradition and unchecked acceptance would doom the haddock fishery.

Herrington's assessment reveals two important points with regard to his arrival as a fisheries biologist near the top of his field in a region with fixed ideas about fishing and fishermen. First, by referring to New England fishermen in the singular, and to that singular character's "conservation nature," he proved his willingness to embrace the popular images of fishing common in New England. He did so, however, to make a point. In this and subsequent papers, Herrington highlighted the radical transformation that New England groundfishing had undergone in a very short period of time. Far from an extension of New England's fishing traditions, Herrington's data described a fishery that was new and modern. Furthermore, Herrington saw that "tradition" had its downsides. Whereas fishermen fully embraced otter trawling across all vessel sizes, they resisted changes to make the gear more selective, and arguably less efficient. Additionally, New Englanders outside the fisheries simply wrote off as "traditional" modern business decisions that condoned the destruction of juvenile fish. Ultimately, Herrington saw that the experiences of New England's haddock fishery needed to lead people to new conclusions: the current mix of tradition and innovation, of present practices camouflaged as perpetuations of the past, had to end. "It is only in the last few years," Herrington informed the American Fisheries Society in 1931, "when the fleet has suffered from a marked scarcity of haddock that the folly of this belief in the inexhaustibility of nature has become potent."[42]

Herrington recognized he needed industry support for his reforms to prove fruitful. Part of that outreach had to be educational. To that end, Herrington and his sometime coauthor, John R. Webster, used the recent crash in haddock landings, which had laid up much of the fleet, to explain to the industry how and why catches fluctuated, and what role fishing played in apparent abundance or scarcity. Writing in the trade journal *Fishing Gazette* in 1933, the two scientists explained the relationship between a given year's spawning success and subsequent years' fish abundance as the smaller fish grew to commercial size: "This increase or decrease in the abundance of haddock as a result of good or poor spawning years does not usually cause an exactly similar increase or decrease in the haddock landings." Landings, in contrast, fluctuated not only with abundance of fish but also as the market provided or removed incentives for fishermen to land them. Those influences affected the number of vessels in the fleet, how hard fishermen fished, and which fish were kept and which were discarded.[43]

More importantly, Herrington and Webster showed that landings, considered in isolation from effort, masked fundamental changes in the health of any given fish stock. Landings data needed to be combined with aggregate data on effort, gear used, numbers of trips taken, and number of vessels fishing to, when collected over time, reveal strikingly different trends from those appearing in landings data alone. "The spawning results were good between 1921 and 1924, causing a great increase in the catch per day during 1926 and 1928. During the period 1925–1928, spawning results decreased to almost nothing and the intensive fishery rapidly reduced the stock of fish on the banks until in 1931 the catch per day was little more than one-quarter as great as during 1927, in spite of improvements in the fishing gear."[44] While landings may have looked good, the amount of time, fuel, and effort needed to achieve those landings had soared during the same period, strongly indicating that fishermen had driven haddock stocks to unprecedented lows.

Herrington also understood the politics behind his findings and presented a remedy unlikely to undermine New England's fishing industry. Herrington believed that changes in the mesh size of the nets used to catch haddock would allow juvenile fish to escape and survive to spawn in the future. In a series of studies using "trouser" trawls, otter trawl nets shaped like a pair of pants with each "leg a different mesh size," Herrington showed that gear modifications could reduce the amount of juvenile haddock wasted in the fishery. He presented preliminary information in 1932, arguing that nets with 5- or 5.25-inch meshes reduced the catch of smaller fish while allowing the capture of larger fish, proved easier to work on deck, reduced purchase price and repair costs, and were easier to pull through the water.[45] To add weight to Herrington's findings, the BCF published his complete results in September 1935.[46] These efforts came to no avail, however. Asked to consider such changes, the industry politely declined to do so and continued to use smaller mesh nets through World War II.

Ultimately, Herrington found in New England a fishery that many celebrated but few fully understood. The New England public still devoured pieces that celebrated fishermen as virtuous cultural icons of a simpler past, as Frederick Wallace had presented in the *National Geographic*. New Englanders found the same romantic appeal in Gloucester's schooner races. And New England's academic writers attached scholarly respectability to these decades-old tropes. Behind this veneer, however, grew an industry built on past profits that fully embraced a capital-intensive, high-volume, mechanized, machine-processed, industrial fishing future. New England's heavy trawling fleet developed in just the eleven years between 1919 and 1930. If the 1919 strikes, the crash of 1929, and schooner races proved indicative, few were ready for modernity in fisheries. Yet culturally, politically, economically, and ecologically, New England's heavy trawlers had created a new world.

The Political Power of Separating Imagery from Actuality

With no more races coming up, Ben Pine and his associates found themselves stuck with *Thebaud*, a schooner too aggressive in design to fish and not aggressive enough to race. But she was a beauty, and Pine knew that the schooner itself had the power to inspire people. In April 1933, he offered *Thebaud* as a solution to "the transportation problem" faced by a contingent of fishing industry leaders from Maine, Gloucester and New Bedford, Massachusetts, and Cape May, New Jersey, heading to Washington, D.C., seeking federal support for the fisheries. In the clearest politicization of schooner nostalgia to date, the 1933 "Sail on Washington" proved to be one of the most significant lobbying campaigns Gloucester's fishing interest had mounted since the 1870s.[47]

On April 6, 1933, the Senate had passed Resolution 15, "Extending Construction Loan Fund Benefits to Fishing and Whaling Industries." If the act became law, fishing vessels would be able to access loans designed to expand merchant marine cargo and passenger carrying capacity. To qualify for the loans, however, newly built vessels would have to be sufficiently large and powerful to meet government service needs in times of national emergency. Within ten days of the bill's passage, Pine decided to send *Thebaud* to Washington to help the bill along.[48]

Thebaud took a delegation of fishermen to Washington to petition Congress for inclusion in Depression-era, federally funded industrial relief programs. Included among the crew were three winning captains of the international schooner races: Marty Welch, Clayton Morrissey, and Ben Pine. From the start, the expedition focused on influencing Congress.[49] Pine originally claimed, with suspicious naiveté for a master at grabbing media attention, "We intend to reach Washington on the morning, state our case and get right out the same day." Such a surgical trip, if that was ever intended, proved elusive, however. *Thebaud* had become a public-relations machine by the time it sailed. After stopping in New Bedford to pick up more delegates, *Thebaud* was joined by a flotilla of fishing vessels from New Jersey and Virginia. Once in the Chesapeake, Essex County, Massachusetts, congressional representative, Gloucester resident, and former Harvard economics professor A. Piatt Andrew arranged for a Coast Guard cutter escort up the Potomac River. The escort drew attention to the flotilla in a critical way: shortly after arriving, President Franklin Delano Roosevelt and British prime minister Ramsey MacDonald, who had been out for a Sunday river sail, came to the waterfront to welcome the fishermen in person.[50]

If *Thebaud* sailed to present a petition from the schooner fleet for New Deal aid, press coverage was more concerned about its embodiment of maritime clichés. Referring to the delegation as "skippers" and "old salts," reporters noted how a New Jersey member arrived "with his duffel bag" before the schooner set sail for the capital. On the ship's arrival in Washington, reporters commented about how *Thebaud*'s crew offered the president a halibut, "and after some scratching of heads, [found] a good salt codfish for Miss Ishbel MacDonald."[51] Indeed, MacDonald's presence was no accident. While some speculated that *Thebaud*'s departure at midnight on April 19 was timed to invoke Paul Revere's famous ride, it is more likely that Congressman Andrew recommended timing their arrival to coincide with that of Canada's premier, R. B. Bennet, the following day. Bennet, with Prime Minister MacDonald, still sought access to American markets for Canadian and British fish. In response, Andrew had arranged for a series of meetings between fishermen and government officials, and likely knew of the premier's impending visit.[52] Despite well-coordinated political planning, the media continued to frame the event as a nostalgic tour of regional characters seeking help for an iconic but dying American industry. Still, while meetings between fishermen and politicians went off pleasantly and even hearteningly, Pine and the rest of the delegation departed Washington uncertain of their fate. It was up to Andrew and his allies to argue before Congress that fishermen were entitled to relief aid.

Thebaud's "Sail on Washington" was not just a publicity stunt, however. As is discussed below, using *Thebaud* as the centerpiece of the Massachusetts delegation's campaign, Andrew spearheaded efforts to include New England's fishing industry in New Deal industrial support programs. Ultimately, success rested on the congressman's ability to coordinate a campaign in the next legislative session that included relief for the fisheries. Working through the winter and the spring of 1934, he oversaw hearings that presented lawmakers with regular doses of romanticism and mythology as they considered the regulation and management of New England's modern industry. Working through allies in the House and Senate, Andrew eventually distilled his initiatives into three bills that sought to include New England fishermen in National Recovery Administration funding.

As a public resource then and now, marine fisheries have been managed via open, if scarcely intelligible, public federal hearings, often hosted by a variety of government bodies that happen to be affecting the marine environment. These hearings, in turn, generated official legislative transcripts that would be consulted in legal disputes. Thus, public hearings, whether before elected officials or agency professionals, become part of the public record and provide unique opportunities for historians. Public hearings allowed supporters to state, on the record, why decision-makers should support any proposed regulation or rulemaking. In doing so, these hearings often also included statements detailing how such requests meshed with American values and mores. Public hearings allowed opponents to make their cases too, forcing discussions and debates over which argument best fit within public expectations of government propriety. As is seen below, such justifications ranged from the factual to the romantic to the histrionic. Critically, however, each of these arguments explains in the public record why each action passed or not. As such, public hearings provided the basis on which any future litigation would be based. In other words, while public meetings may have been choreographed to facilitate the committee's approving the outcomes the main parties had agreed to ahead of time, public hearings were no mere formalities. How industry leaders framed their requests ultimately reflected how they believed their goals best meshed with contemporary American legal, political, social, and even cultural conventions. Furthermore, throughout public hearings, lobbyists sought to justify public support for private industry by invoking the rationales they felt appealed most to the larger public. Often such rationales rested on the same imagery that had long characterized New England's popular perception of its fishery: quaint, romantic, traditional, and timeless. Such imagery provided too good a tool for lobbyists to ignore and too good a political mantle for politician to refuse. In all cases, however, public hearings represented, and still represent, unique moments when what was asked was framed explicitly in why it should or should not be granted.

This certainly was the case as Andrew and other representatives from New England worked to secure federal aid for New England fishermen. As the nation plummeted further into the Great Depression during the year after the Sail on Washington, Americans wrestled with limiting government interference in markets and industries while at the same time recognizing that only federal action could prevent the economic collapse from getting worse. Issues of how federal efforts would prop up the economy, and who would receive that support, posed unique challenges to accepted American traditions. For Andrew, who sought to aid fisheries by means of industrial support programs, *Thebaud*'s Sail on

Washington provided the perfect opportunity to publicize the important role that fisheries, like other industries, had played in America's economic development.

In the first of three legislative campaigns launched in 1934, Andrew sought to secure merchant vessel construction credit and loan guarantees for fishing vessels, effectively slashing fishermen's cost of upgrading operations. For this initiative to pass into law, however, Andrew had to secure the blessing of the House subcommittee tasked with evaluating in close detail matters pertaining to fisheries. Far from a de facto favorable audience, this committee included those who would support Andrew's proposal and those who would oppose it. Therefore, when public hearings opened on the bill before the House Committee on Merchant Marine, Radio, and Fisheries, the Massachusetts congressman appealed to his colleagues to "consider themselves the voice of a small industry, which is an important industry and the oldest industry in the country, but which today has very little voice in the government." It was only fair, Andrew believed, that an industry so important to American history, heritage, values, and its Anglo-American past, should be eligible for federal aid during times of need.[53]

The fishing industry loved the idea, but even the Sail on Washington could not sway all the support the bill needed. Congressmen George Edmonds of Pennsylvania and Frederick Lehlbach of New Jersey both challenged Gloucester's claim to the mantle of tradition. As Andrew explained to the committee, Gloucester hosted "no very large companies, no great corporations, engaged in the fishing industry. In my town of Gloucester I should say there were few firms that had more than four or five fishing vessels." Edmonds pressed back, asking whether the port's large mackerel and herring vessels fishing off Cape May were "corporation boats" owned by large firms operating out of Gloucester. Lehlbach eschewed subtlety altogether by recalling that one of the nation's largest fishing firms sat squarely in the middle of Gloucester: "Take a concern like the Gorton concern—that's in your town, isn't it?" "Yes," Andrew replied. "They are not so much a producing firm. . . . They have [only] a few vessels."[54]

Lehlbach did not let up, however. Next he challenged the fisheries' ethnic claims. "The fishing industry is largely in the hands of the Portuguese, is it not?" Andrew countered: "You mean Portuguese-Americans?" In expanding his response, Andrew's argument for fishermen's ethnic and national importance began to unravel: "Well, up there originally they were largely Nova Scotians and some Scandinavians. But the sons of the old fishermen do not care to pursue the calling; they would rather sell automobiles or pursue softer perhaps more lucrative callings, and so the fishing business, with each generation, changes its personnel. There are not many sons of fishermen who engage in that calling." Catching his weakened argument, however, Andrew added, "Now there are a good many Portuguese who came from the Azores and some Italians who came from Sicily, and there are also still a great number of the old Nova Scotians and some Scandinavians."[55]

It was a notable blunder that required remediation by Representative John W. McCormack of Boston. McCormack succinctly reminded the committee of the industry's historical and modern importance. For McCormack, the Boston Fish Pier within his district was "the greatest scene of activity in fishing in New England and one of the greatest in the United States, if not the world." Notwithstanding such activity, history and scale had not insulated the Boston Fish Pier from contemporary trends: "The fishing industry, like other industries, felt the effect of the depression." They were trying, like all others, "to bring back into operation the law of supply and demand, thereby [to] revive private industry and reabsorb the displaced workers." Furthermore, McCormack continued, federal relief to businesses stood as "one of the theories upon which we are compelled to resort to the use of the power and influence of the Government in this emergency, necessary as a result of the breakdown of private industry." McCormack's argument for including fishing vessels in relief programs rested on understandings of the relationship between industry and government, between public funds and private property, and between employers and employees that had evolved through American industrialization. But McCormack also mobilized history to support his case. Briefly summarizing Samuel Eliot Morison's conclusions published the previous decade, McCormack cited the role of "those hardy seamen" in colonial settlement and as the initial spark for American independence. He also credited fishermen for winning the revolution itself, adding that "some historians give the credit [to New

England fishermen] for finally compelling England to enter in to a treaty of peace."⁵⁶

Other committee members quickly pressed McCormack on the contemporaneity of such traditions. David D. Terry from Arkansas asked, "Is it true, as stated by Mr. Andrew, that the sons do not follow the fathers in the fishing industry?" McCormack had to agree. "I assume in these modern days it is—that that is true." But it was also true for many other forms of work, McCormack countered: "The farm has become more or less industrialized, where at one time it was a home, not even an investment—it was a home." Still, McCormack had to concede that the preceding half-century had changed New England's oldest industries: "I should say that is true, . . . that the old tradition, family tradition, does not exist. Although in part it probably does; yet, generally speaking, it has disappeared."⁵⁷ In the end, committee questions forced McCormack to speak on behalf of fisheries both past and present: "I respectfully submit that this ancient but always new, this old but always essential and basic industry, which has contributed so much towards the progress of our country, is entitled to the same emergency considerations that have been extended to other business activities."⁵⁸

If extending vessel construction loans to the fishing industry challenged Andrew and McCormack to portray the fisheries as both modern and traditional, hearings on Andrew's second of three bills, H.R. 9015, turned out to be far easier. In this bill, again presented before the House Committee on Merchant Marine, Radio, and Fisheries, Andrew called for extending a variety of agricultural relief benefits enjoyed by America's individual farmers to New England's individual fishermen—and in particular, the small-scale fishing operations that existed alongside the larger firms benefiting from vessel construction loan guarantees. These debates, far more than his first bill, afforded Andrew the opportunity to speak almost exclusively to the plight of New England's independent, small-scale fishermen. In doing so, he presented the committee with some of the most evocative images and symbols of the fisheries' past.

In these hearings, Andrew's strategy revealed how powerful fishermen's iconography played outside New England. The tone set by the committee chair, Virginia representative Schuyler O. Bland, reveals Andrew's successful use of this imagery to secure political support from unlikely allies. In opening the committee hearings, Bland clearly established the proceeding's overall tone by stating that New England's fisheries were "made up of an individualistic group," but one whose traditions have prevented them from organizing and advocating for government support. Andrew built upon this imagery, claiming that if passed, the bill would redress a social as well as an industrial wrong. Ignoring the significant sums spent on researching, developing, and enhancing New England's fisheries over the preceding decades, Andrew claimed, "[The bill] will only bring about justice so far as a neglected industry is concerned. The fishermen are really the 'forgotten men.'" Former Gloucester mayor William J. MacInnis supported this theme: "The fishermen are individualists," he stated. Fishing's very nature rendered it inevitable that individualists would always be drawn to fishing. "It is a free life, a hard, a difficult life, but it is free." Notwithstanding the consolidation of Gloucester's schooner fleets before World War I, and ignoring the more recent expansion of the General Seafoods Corporation in the city, MacInnis claimed, "For that reason, they will be individualists and will never probably get together into any few large corporations like the other great industries, for instance the automobile industry." MacInnis further testified, "Every fisherman aspires to have his own boat, to have his own rig," echoing characterizations from Morison's histories, as McCormack had done earlier. Nor did MacInnis overlook earlier arguments about fishermen's role in American history: "I think many times we are inclined to overlook the great part that the fisheries have played in the history of our country, particularly in war time.... I think it was our great Senator, Daniel Webster, who said that the fishermen and the farmers are the founders of civilization."[59]

Framing the bill in human, social, cultural, and ideological terms made good political sense. Indeed, blurring the line between popular perceptions of New England fishing and its industrial nature had been a reliable tactic since the 1890s. By the 1930s, with national attention piqued through the international schooner races and *Thebaud*'s Sail on Washington, nostalgic, broad-stroke characterizations of New England's fishing heritage had begun to seep into harder economic analyses and regulatory debates affecting industrial reality.

Compared to the fanfare and media coverage of the hearings surrounding their first two bills, hearings for the Massachusetts delegation's third bill to secure federal aid for fishermen was a subdued affair. In a brief one-hour hearing, H.R. 8930 called for the appropriation of funds to design and construct a fisheries research vessel to replace the aging *Albatross II*, a retrofitted navy tug constructed in 1906 and commissioned into fisheries service in 1926. Such a vessel would allow the United States to extend fisheries research operations into winter months, something *Albatross II* could not do safely. Expanded seasonal operation would allow more comprehensive research to improve fisheries development and regulation.[60]

Taken as a whole, Andrew's campaign reveals how blurred lines had become between popular fisheries imagery and its industrial reality. The first bill sought federal support to expand, industrialize, and mechanize New England fishing, but the second bill revealed how American sentimentalism and romanticism confused fact and fiction in how the fishery would be regulated. The third hearing immediately acknowledged the industrial expansion and modernization that had defined the industry since 1919, and the bill made clear the consequences of the first two. In laying out the case for the new vessel, BCF's division chief of scientific inquiry, Elmer Higgins, explained to the committee in no uncertain terms that the rapid expansion of the haddock fisheries over previous decades threatened its future viability.

Herrington's haddock work provided Higgins with a firm foundation to justify the new research vessel. Acknowledging recent economic forces that were driving down landings, Higgins argued nevertheless that "most of the ills of the fisheries may be traced to lack of knowledge concerning the fluctuations in abundance of fish. . . . Suitable data collected by a research vessel could give the clue to impending changes." More important, Higgins continued, was the role of good data in preventing overexploitation. "Further, the instability of abundance even when its causes are known, makes it difficult to perceive what the potential average yield of the resources may be and whether or not the toll exacted by the fishery making its inroads beyond the capability of nature's reproductive forces to replace." He looked to the haddock fisheries for evidence: "The overexpansion of the trawling fleet during the boom of abundance of haddock in 1927 and 1928, and the tying up of the vessels during the subsequent decline in abundance, aggravated by the depression, is an example of the cost of ignorance." Higgins was not alone in these views. The board of directors of the Federated Fishing Boats of New England and New York, a body representing most of the Northeast's large trawling firms, "heartily" endorsed previous research conducted by the USBF and BCF and "earnestly" urged the committee to support the bill.[61]

Additionally, a member of the New England haddock trawling fleet, John Graham, submitted a manuscript article, then in preparation for publication in the industry journal *Fishing Gazette*, titled "The Haddock and Cod Fishery Need a Research Vessel." Graham's piece revealed that while politicians and lobbyists invoked heritage and tradition, industry members looked to analytical tools emerging from the scientific work of E. S. Russell, Michael Graham, William F. Thompson, and William C. Herrington. For two and a half decades of growth, Graham wrote, the New England trawler fleet "was never seriously threatened by the scarcity of fish. Are we to consider the decrease in per boat per day production which occurred from 1928 to 1933 as unusual, or are we to take it as an indication of what we may expect of the future?" Graham continued, "We should be blind and stubborn were we to ignore the history of fisheries older than ours." Considering fluctuations in the fisheries in Great Britain, France, and Newfoundland, "we must inevitably conclude that we are faced with much the same problems. We have no basis for the assumption that we shall always enjoy prosperous fishing and that the years 1928 through 1933 were abnormal and exceptional." Converted governmental research on physical, biological, and industrial influences could address these questions and, most importantly, keep the fishery from destroying itself.[62]

Massachusetts's legislative campaign in the spring of 1934 provided a rich and exhaustive demonstration of how publicity, perception, and politics could be blended to effect regulatory and legislative goals. The hearings also revealed how powerful and contested the stock tropes of New England fishermen had become. While some rejected Andrew's and McCormack's romantic imagery, others embraced it. Before the end of the congressional session in June 1934, Boston's large trawling firms qualified for vessel construction loans, and Congress had approved the construction of a new fisheries research vessel. Both of these bills focused on the modern aspects of New England's fishing industry: Boston's heavy trawlers and the role that science would play in optimizing their catches. Ironically, the bill that would have supported the most iconic of the region's fishermen, the small-scale operators sailing out of Gloucester, did not pass—but not because Andrew's efforts were unsuccessful. The bill allowing fishermen to benefit from agricultural subsidies, H.R. 9015, "For the Relief of Persons Engaged in the Fishing Industry," passed both the House and Senate at the close of the legislative session. But President Franklin Roosevelt, a Democrat, pocket-vetoed this bill, which would have helped Andrew's largely Republican district.

From *Thebaud*'s Sail on Washington to its multipronged legislative push to win government support for New England's fisheries, Massachusetts's representatives waged impressive campaigns in 1933 and 1934. As savvy legislators, Andrew and McCormack used all the tools at their disposal, securing important allies in key committee positions, employing visually compelling publicity, offering cultural arguments that appealed to the heart and economic arguments that appealed to the head. Quite simply, it made good political sense to blur the differences between the fisheries' romantic past and modern present, but one must also bury the consequences that resulted. Like Gloucester's representatives in the 1870s and Charles Francis Chamberlayne in the 1890s, Andrew and McCormack proclaimed the New England fisheries' virtuous heritage and historical tradition. But they showed that the fisheries were modern enough to deserve the same federal support that farming and manufacturing received under the New Deal. In this manner, the 1934 campaign exposed an important tension that confounded even the most ardent supporters of New England fisheries: beneath the lament of tradition stood a modern fishery that operated just like any other modern industry. How and when to use a traditional image or one of modernity would shape New England's fisheries lobbying for the next thirty years.

Myth and Reality on the Eve of World War II

In May 1937, Metro-Goldwyn-Mayer released the film adaptation of Rudyard Kipling's *Captains Courageous*. With an all-star cast, the film presented in live action a book whose American sales had made it enormously successful. The film also proved lucrative. Released to cinemas in the dying gasp of excitement surrounding international schooner racing, the film appeared as prolonged economic hardships sent Americans to the movies seeking escape, excitement, solace, and perhaps guidance as to what the country actually meant. For viewers, the film reaffirmed the surety and transformative power of American values. Its star, Spencer Tracy, playing the Portuguese doryman Manuel, won an Oscar for best actor.

Reviewers identified the portrayal of the Gloucester schooner fleet as one of the film's high points. Frank S. Nugent, writing for the *New York Times*, imbued the film with the power to represent contemporary fishing, which the story had eschewed almost forty years earlier at its original publication. In his critical review, Nugent wrote, "The picture does not really come alive until the cameras turn upon the [schooner] *We're Here*. . . . Then, in its depiction of the men and methods of the old Gloucester fleet, it takes on almost the quality of a documentary film, enriched by the poetic photography of schooners spanking along under full sail, of dories being lowered into a running sea, and shading in, quite deftly, the human portraits of the fishermen with their quiet heroism and resignation, their Down East humor, and their stern code of decency."[63] After almost seventy years, the ageless image of New England fishermen had come to the silver screen, and Americans embraced its characterizations as substitutes for the industry's reality.

At the same time, a decade and a half of intense industrialization and mechanization left others unwilling to accept old tropes. A young geography instructor at Harvard in the late 1930s, Edward A. Ackerman, broke ranks with his elder colleagues, who had long embraced and sustained the old motif, and took the industry to task for the biological consequences of its actions. Publishing in the journal *Economic Geography*, Ackerman issued a harsh critique of the industry's failure to plan ahead as to how it would harvest current fish stocks sustainably: "No plan exists, no consideration has been taken for the important marine fisheries. Like the forests of yesterday, the fisheries of today in most cases are assumed to be self-perpetuating." Ironically, Ackerman noted, for an industry that had embraced history and tradition in recent decades, the absence of management reflected a willful historical ignorance: "The lessons of the past are forgotten, or at least they are not considered applicable to present-day fisheries." Instead, Ackerman took his New England neighbors to task, along with the fishing industries, in their willful denial of historical patterns. He noted, "The story of the virtual extinction of the whale, salmon, and shad are familiar enough to many New Englanders, but few of them can imagine that the now abundant haddock, flounder, and redfish may have a similar fate."[64]

When viewed comparatively and in broad strokes, Ackerman argued, a clear cycle emerges from New England's fisheries that spelled doom for the local fish stocks on which the industry relied: "As soon as a good market appeared for a form of life in the sea it was pursued relentlessly until its scarcity made protection imperative, or fishing no longer profitable." Then, using examples from the region's salmon, shad, whale, halibut, mackerel, and menhaden fisheries—almost all fisheries actively pursued in the nineteenth century besides the cod fishery—Ackerman called for meaningful restrictions on gear and fishing pressure to ensure that cod, haddock, lobster, flounder, and redfish, would not suffer the same fate. He argued, "One can illustrate the trend with stories of other depletions (hake for example), but all such stories emphasize the same point: In New England waters no regulation, no limitation is undertaken until a fishery is not merely threatened, but seriously impaired. What is happening today is more significant. At this moment New England is fishing out haddock, flounder, and redfish more intensively than it ever fished lobster, salmon, and shad." To Ackerman the continuation of the pattern was a disgrace: "We think back with contempt for the depletions which earlier generations brought on, yet we go on quite blind to the effects of our present activity."[65] Fishing could no longer continue as usual; some forward thinking needed to happen if the industry was to remain viable and escape past patterns of boom and bust.

Taken together, the transposition of past and present in *Captains Courageous* and Ackerman's scathing indictment of New England's failure to regulate its fisheries reflect an uncertainty about the industry on the eve of World War II. Following the scandals surrounding the New England Fish Exchange, the region and its academics embraced fisheries heritage in a nostalgic binge lasting over two decades. Modern, mechanized, industrial fisheries continued to peak out from behind that veil, however, disrupting a collective desire, or an unconsidered inclination, to ignore modern realities. For the industry, the confusion worked just fine. If legislators felt that extending vessel construction loan programs to the fisheries or paying for a fisheries research vessel helped the iconic, humble, hearty, brave, and stoic small-operator fisherman, so be it. However, work by Herrington and later Ackerman could not easily be dismissed. Fishermen everywhere saw with their own eyes that the standing stock of haddock, which had fattened Boston's fisheries, was no more. Calls for regulation, such as Herrington's recommendations for larger mesh sizes, ignored by the industry as late as 1936, would have to be heeded sooner or later.[66] But questions remained as to how regulations could be enacted without harming New England's oldest and most virtuous industry. War, and redefining New England's fishing "traditions," would provide some answers.

CHAPTER 4

REINVENTING TRADITION

New England Embraces Industrial Fishing

While New England's politicians invoked tradition in Washington, D.C., most Depression-era New Englanders found that tradition would not solve their problems. With large numbers out of work and hungry, debates over what New England's fishery *should* look like appeared quaint and irrelevant. Grim necessity demanded that modernity trump nostalgia as the industry struggled to land food for hungry people cheaply and efficiently. Exigencies of the Depression and later of World War II helped people accept the fisheries' newfangled modernity and efficiency. Fishermen might not handle sails or fish from dories, yet workers in steam- and diesel-powered fishing vessels could still land food. As a result, New England's cultural arbiters soon embraced the industry's mechanized form, now over a decade old, as the new definition of "traditional." As steam- and diesel-powered vessels became part of evolving fishing traditions, so too did the industrial organizations at the core of New England's fisheries. Two organizations came to highlight how modern New England fishing had become. Out of Boston's heavy trawler haddock fishery emerged a formidable labor union and an equally formidable owners' association. Gloucester's fleet and processors also organized, and New Bedford's eventually followed. In the face of hard times and then war, few mourned the passing of the schooners, dories, and baited hooks. Given the dire circumstances of the decade, New Englanders embraced and even celebrated, albeit for a limited time, the modern industry that decades ago had come to define commercial fishing.

The Depression and Increased Pressure on Fish and Fishermen

The Depression fundamentally changed the nature of New England fishing. With falling commodity prices and lower yields compared to booming times in the mid-1920s, gloom and inactivity settled on Boston's heavy trawler fleet. Lost disposable income reduced the demand for food in many working families, which in turn depressed revenue in a bleak negative feedback loop. To prop up revenues, fishermen fished harder making matters worse. According to R. H. Fiedler's annual reports for the U.S. Commissioner of Fisheries between 1931 and 1938, New England's heavy trawlers (91 tons and larger, based mostly out of Boston) increased the total trips taken each year from roughly 1,250 to more than 2,500. Similarly, the cumulative number of days they spent at sea per year also increased from roughly 10,000 in 1932 to over 16,000 in 1938, down from the 1937 high of more than 18,000 days at sea.[1]

Fishing efforts remained close to home, however. The average voyage length of each large otter trawler fell markedly during the 1930s, suggesting that captains took shorter trips and targeted local stocks, likely to save fuel costs. In 1931, the average trip for a large otter trawler lasted roughly ten days. After increasing to about twelve days in 1934 and 1935, trip lengths fell to only six days from 1936 to 1938. While shorter trips could still target eastern Georges Bank and even Browns Bank (lying fifty miles south of Nova Scotia's southern end), the vast majority of larger otter trawl trips continued to exploit haddock stocks on western Georges and the Great South Channel. Good yields close to home became a recipe for survival for the sixty most active, very large heavy trawlers, which consistently accounted for a third of all edible fish landed in Boston, Portland, and Gloucester.[2]

Smaller vessels also shifted fishing strategies. In the small otter trawl fleet (vessels between five and twenty tons net burthen), operators increased the number of trips and expanded the total number of days they fished over the year. The number of active vessels expanded as smaller boats offered operational savings as well as access to inshore stocks. From 1932 to 1938, the small otter trawling fleet expanded from around forty five vessels to just under one hundred; they quadrupled their number of fishing trips (from 500 to 2,000 trips per year) and increased days absent from port by more than 350 percent (from roughly 1,800 to just over 7,000 days per year).

Midsized otter trawlers (between twenty-one and ninety tons) before the Depression were squeezed between the expanding fleet of heavy trawlers, which landed larger fares, and the small otter trawlers, which cost less to operate. Between 1929 and 1932, the midsized fleet shrank from roughly 140 vessels to about 65. It peaked again at 100 trawlers in 1937 but lost 10 vessels the following year. Numbers of trips and days absent from port declined in 1932, then grew modestly compared to the other vessel classes until 1938.

From the fishing community's perspective, the Depression intensified work routines dramatically. While government officials and businessmen paid attention to comings and goings from port, the real measure of work for fishermen in Boston, Gloucester, and Portland was how long each fisherman spent fishing. Even short trips that targeted coastal stocks kept fishermen away for whole days at a time. Consequently, how often fishermen were absent from home governed how the rest of their community gauged changing fishing patterns. For example, the 180 days per year each heavy otter trawler spent at sea in 1931 must have seemed positively leisurely compared to the peak year of 1936, when large otter trawlers and their crews averaged 300 days at sea. With such an intense schedule, vessels could have alternated crews, which would have spread employment among unionized fishermen, but the records are silent on this issue. By 1939, however, sea time had declined to 250 days per year, with the very real possibility that the intense work—between eight and ten months at sea per year—over-strained crews, their families, and communities. In comparison, small otter trawling crews had it much easier. While the annual number of days they spent at sea tripled, the 1939 peak saw each vessel averaging only 75 days at sea a year. Medium-sized dragger crews actually saw their sea time shrink from roughly 120 days absent per vessel per year to just under 100. As the fleet expanded, so did the intensity of the work. Together, larger fleets and harder work routines turned up the dial on fishing in the 1930s, as New England's fleets bore down on local stocks of fish.

From Traditional to Modern: The Cultural Redefinition of New England Fishing

Increasing intensity of the work of fishing altered the way New England's cultural commentators portrayed the industry. The last thing modern, mechanized fishing firms wanted was to be seen as quaint, timeless, and traditional. Eager to attract investors, many of Boston's large firms preferred instead to present themselves as cutting-edge innovators. As early as 1931, Frank H. Wood, writing for the reorganized Bay State Fishing Company, published a history of their flagship prepackaged, frozen haddock fillets, "40-Fathom Fish." From the start, the firm defined itself as the next evolution of New England's traditional industry. With the opening section titled "An Old Industry Is Re-Born," Wood both embraced the heroic imagery of New England's fishing past and extolled the modern trawlers, which offered a safer and more efficient fishery: "Until the Bay State Fishing Company introduced the modern method of fishing known as trawling, the fisherman's calling was indeed dangerous and toilsome." Self-serving corporate promotional literature, Wood's story continued to tout innovation and enhanced workplace safety. "Partly to avoid the dangers of dory fishing and partly to develop a new method of fishing that would be certain to supply fresh fish regularly," Wood continued, "the Bay State Fishing Company was organized in 1905 to build and operate the first trawler in American waters." Certainly, profitability and market share influenced this decision as well. Wood adroitly evaded the inconvenient matter of the 1919 indictments. Rather than acknowledge the company's convictions on war profiteering and monopolizing the fishing industry, Wood characterized the event as the failure of previous management to secure sufficient markets. Regardless, Wood did not want past embarrassments to detract from the firm's accomplishments: "To the Bay State must [go] credit for the first awakening [of New England's fishing potential], and most of the advances since have originated with this Company."[3]

Wood's pamphlet was more advertising than history, written to appeal to customers, potential retailers, and possible investors. Therein lies the work's importance. In representing itself to the larger public, Bay State's marketing executives chose to embrace an industrial identity of modernity, rebirth, and "awakening" that distanced it from previous images of New England fishing's timelessness, tradition, and antimodern symbolism. While Gloucester and the rest of the country look backward at the international schooner races, Wood and Bay State stared forward and pulled all of Boston's fleet with them.

Others shared Wood's vision, and by the late 1930s writers outside corporate halls presented similar images. By the mid- and late 1930s few freelance writers and mass-market publishers bemoaned the loss of the schooner fleet, in stark contrast with attitudes less than a decade earlier. To writer Bernard Breedlove, who described the industry in the *Saturday Evening Post* in September 1938, the otter trawler fleet represented the sheer power of modern technology: "The engine-room telegraph jingles: a muffled explosion rumbles from the vessel's hot Diesel exhaust, and the 750-horsepower engine throbs, takes hold and kicks astern. . . . A modern fishing vessel has sailed for the North Atlantic banks." In addition to engine power, Breedlove highlighted how cutting-edge communications technologies in trawlers greatly enhanced their capacity to quickly and efficiently harvest fish: "The trawler radio operator, handling wireless equipment equal in efficiency and power to that installed on the large ocean-liners, may reach as many as fifteen fishing vessels in as many minutes." The geographic footprint of radio communications took much of the guesswork out of fishing. Breedlove observed, "In that time radio covers nearly 100,000 square miles of fishing territory, and gives the trawler captain the equivalent of days of trial-and-error steaming and fishing over the banks." While no substitute for skill, the captain's experience and up-to-date intelligence allowed him to quickly determine the potential for a speedy catch and a full hold. Modern vessels also benefited from "a device known as a fathometer. This is an electrical depth-finder, operating on the principles of radio . . . [and is] one of the greatest of modern aids to the fisherman." Between diesel engines, radio communications, and electronic navigational equipment, Breedlove celebrated how Boston's modern trawler fleet could maintain a fully industrial production schedule: "On and on, night and day, around the clock the fishermen work in two shifts, six hours on deck and six below. . . . As soon as the vessel is loaded with a catch that will merit putting in to the Boston Fish Exchange, orders are received by radio from the home offices and the course is set for home."[4] Far from the romantic vision depicting New England's fisheries a few years earlier, the *Saturday Evening Post*'s nationwide readership received a picture of modern industrial efficiency, technology, and power at work in the mass production of fish products for home consumption.

By 1941, the success of otter trawlers in landing food for a hungry workforce led even Edward Ackerman to embrace New England's modern fishing industry. Stepping back from his 1938 indictment of the serial depletions caused by New England fishing, Ackerman's book-length study, *New England's Fishing Industry* (1941), marked a key transition. First, he acknowledged the romantic rhetoric of the past as he analyzed the structure and function of New England's contemporary fishing industry. Then he utterly dismissed it, conceding that efficient gear, modern power plants, and integrated shoreside processing and distribution had forever transformed a traditional industry into a progressive one.

Ackerman continued to embrace, to some degree, earlier rhetoric of fishermen as forgotten men. "Only a generation unconcerned with history and tradition," he wrote, "could so far forget the foundations of New England's greatness." While mills and manufacturing had played a prominent role of late, "modern New Englanders are not likely to remember how much they owe their neighboring sea." And in what by the late 1930s had become a historical trope, Ackerman briefly recounted the conventional New England story of fish and colonial settlement. However, as a student of the present as well as the past, he clearly saw that regional romanticization masked industrial modernity and efficiency: "The romantic days of sailing ships and Captains Courageous [sic] are gone as completely as the New Bedford whalers and the Gloucester salt halibut fleet." In their place, new gear and new propulsion marked the modern fishery: "Nets, seines, and trawls have replaced much of the old hook-and-line fishing, and no fisherman in New England now depends on the wind for his motive power. . . . Even the days of steam have passed. Diesel and gasoline engines turn propellors wherever fish are sought, and they helped to lighten the actual labor of fishing enormously."[5]

It was in Ackerman's revised views of the environmental impacts of modern gear that his tune changed most dramatically, however. Just three years earlier, Ackerman had condemned New England fishing industries for their tendency to deplete stock after stock of valuable species, and in doing so, made a clear case for restrictive fishing regulations. His 1941 study still raised serious questions about otter trawling's effects on juvenile fish. He wrote, "The widespread use of the otter trawl may have been unfortunate," because it caught everything in its path and forced fishermen to sort through both marketable and unmarketable species with every haul-back. As a result, otter trawls caught young, submarket-sized commercial species, which then required wasteful discarding. "Throwing the small [cod or haddock] overboard does no good, because the pressure of the catch in the a towed net kills many individuals in it, and the inflation of [fish's] air bladders occasioned by being brought to the surface prevents the remainder from regaining the bottom." Ackerman also cited the 1913 USBF study that documented that 22.4 percent of all haddock and 19.2 percent of all cod taken in otter trawls were below market size and tossed overboard.[6]

Otter trawling's effect on the benthic habitat also raised concerns: "Various fishermen insist that [an otter trawl] sweeps the bottom so clean in its passage that it destroys the feeding grounds of the fish." Indeed, this claim was made by hook fishermen in the 1912 Gardner bill hearings. Thirty years later, the "few fishermen who still do hand lining are loudest in their denunciations," and even some otter trawlers conceded (as they had done in 1912) that "trawling leaves the bottom in poor condition." While Ackerman acknowledged that the 1913 studies refuted such claims, fraught as they were with shifting goals and changing terms of reference, he continued to argue that "repeated dragging often has caused small inshore grounds to become entirely depopulated of fish. This is notable on some of the grounds off Cape Cod." Alarmingly, this proved true for "the flounder grounds around the mouth of Narragansett Bay," a fishery that had been exploited for two decades at most.[7]

Despite these misgivings, Ackerman backed away from his earlier firm stance on fishing and depletion. Instead, he presented a much more ambiguous and uncertain set of conclusions regarding otter trawling and long-term sustainability: "Although it is certain that the unlimited use of the otter trawl could deplete fishing grounds, the extent to which trawling should be permitted can be decided only with many more years of experimentation and detailed experienced observation of the fisheries." Drawing from turn of the century scientific studies, Ackerman suggested that otter trawling might even enhance fishery productivity: "A reduction in the number of large fish may mean a higher rate of survival of young because of decreased competition for food."[8]

Regardless of the gear used, Ackerman came to view fishing's long-term sustainability as a political problem. "When the fish seem to be maintaining their own," he observed, "legal restrictions are little thought of; but when fishermen have upset the balance too far for their own or their community's good, political limitations must be set."[9] The prospect of such regulation politicized the fishery: "It is impossible, however, to escape the suggestion that for its existence the New England fishing industry in general is dependent upon political, as well as on physical, factors. This was well proved in the agitation which attended the recent (1939) reduction in duties on imports of Canadian fish."[10]

While softening his argument that otter trawling would inevitably lead to fish stock depletion, Ackerman, more than any other observer or researcher, saw that fisheries were now a modern industry. Like other heavy, extractive industries, the industry was shaped both by political influences and by the natural resources on which it relied. "As in the case of fishing methods," Ackerman wrote, "it is difficult to decide how much political and legal facts are results of fishing activities and how much they are elements in determining its character. On the one hand, they are certainly a part of the cultural picture . . . on the other hand, although they are far from immutable, these elements of the background of commercial fisheries are just as real for the life of fishermen as tides and storms . . . and the migrations of fish." Although the physical and biological environment shaped how much fish the industry could take, business and politics defined the rest of the catching and selling equation: "The banks fishery depends in part for its existence on the laws which keep its market more or less exclusive [from foreign competition]."[11] While New England fishing firms sought to be the modern manifestation of New England's traditional industry, Ackerman's observations on the role of politics in the productive process highlighted how modern the industry had become.

As the industry shed its traditional mantle and cultural observers welcomed its modernity, Ackerman's study was the first to identify how New England's major fishing industries modernized not only mechanically but also politically. He identified the important intersection of business and regulation, which helped bring the modern industry into existence. An aggressively protectionist tariff policy against Canadian imports, combined with a lax, laissez-faire domestic regulatory regime, provided a fertile commercial climate in which fishing could thrive.[12] For an outside observer, these stark findings marked a dramatic change from previous examinations more concerned with confirming mythic heritage. Ironically, although Ackerman's perspective may have been new to industry outsiders, New England fisheries' political engagement, both domestic and international, dated back almost a century. With modern equipment, seemingly abundant stocks, and a thriving wartime market, these realities were finally ready for full public view.

The "New" New England Fisherman

New England's industrial fishing operations modernized in ways familiar to other contemporary American industries. While Boston's processors, wholesalers, and large fleet owners celebrated the technological sophistication of catching, processing, and distributing fish products, fishermen working on the heavy trawlers modernized as well. In the fall of 1934, three thousand food processors, fishermen, and dock workers, represented by various marine trade unions under the auspices of a new labor organization called the American Fisheries Union (AFU, no relation to the Gloucester body from the 1880s), walked out on strike at Boston Fish Pier. The strike soon attracted support from crews of incoming vessels and unions in New Bedford and sparked the formation of an entirely new union in Gloucester, which promptly joined in the action.[13]

Employers responded in creative ways that ignored the labor unrest. Rather than confront strikers with scab labor or police, the General Seafoods Corporation used the Boston Food Fair across town to present an alternative narrative to the public. Even as fishermen, processors, and dockworkers picketed the docks, "heavily" according to one account, General Seafoods omitted entirely the people who transformed wild organisms into commodities, presenting instead the natural world of fish and ocean banks. "Large maps and photographs show the water[s] which supply the country with fresh fish," recounted one reviewer. Alongside the display, sixteen dealers arrayed before the viewing public a "splendid 'catch'" of select cod, haddock, scrod haddock, butterfish, catfish, mackerel, cusk, smelts, grey sole, lemon sole, halibut, herring, flounders, pollock, salmon, live lobsters, and a slew of other shellfish. Notably missing from the exhibition were fishermen themselves. In industrialists' idealized vision of New England's modern fishing, unionized fishermen had no place.[14]

In reality, however, the strike ground on throughout October, laying up a considerable proportion of the Boston fleet and significantly decreasing daily landings. The New Bedford chapter of the Fishermen's Union of the Atlantic completely tied up that fleet, and by October 23, senior union leaders headed there to hammer out an end to the action. Although New Bedford fishermen did not want to interfere with the Boston strike, their willingness to cut a deal undermined the resolve of strikers in the rest of the region. By the end of October, ten trawlers sailed out of Boston crewed by union and nonunion fishermen: the strike had effectively ended.[15]

Reorganized under a new union with new leadership, Boston's fishermen fared better a few years later. The AFU threatened to walk out on March 1, 1939, after fifteen months of failed contract negotiations. In addition to concessions on wages and work conditions, fishermen also sought to limit the catches of certain fish to 125,000 pounds (56.7 mt) per trip to prevent glutting markets, which depressed auction prices and fishermen's earnings. With most large trawlers capable of handling 400,000 pounds (181 mt) of fish per trip, these limits threatened to drive up prices, reduce volume, and undercut vessel profitability.[16]

An eleventh hour deal averted the strike, which would have tied up roughly one hundred large trawlers in Boston and Gloucester. Still, the deal marked a tremendous win for the AFU's new secretary, a Newfoundlander named Patrick J. McHugh. With charismatic leadership and trade unionist ideals, McHugh leveraged a new, one-year contract that granted fishermen significant improvements in wages, working conditions, and working schedules. In addition, under the new contract, trawler fishing became a closed shop and crews would receive a better split of the profits. Owners also agreed to establish a grievance board, guarantee a twenty-five-dollar-per-day minimum wage in case no fish were landed, and standardize vessel departures with minimum dockside layovers in case of repairs. Crews would also not have to fish on holidays. In exchange for concessions that brought New England trawler fishermen's work more in line with shoreside labor practices, fishermen dropped other demands for owners to cover more overhead costs. Crucially, fishermen also dropped their demands for catch limits.[17]

The AFU's success added a unionized workforce to New England fisheries' march to the modern. Like other workers, fishermen sought stabilized working routines, holidays off, and a minimum wage in case of poor trips. They also adopted modern labor management and representation mechanisms, and in exchange, ownership streamlined contract negotiations across the entire fleet. Besides mechanization and politicization, the 1939 contract marked one more way in which Boston's and Gloucester's large vessel fisheries had modernized: blending together the shared risk and reward of traditional pay arrangements—the share system—with the safety net of industrial wage structures.

Despite these concessions and efficiencies, or perhaps because of them, tensions between large trawler owners and the AFU flared regularly over the next decade, often making front-page news. In 1940, owners locked out fishermen in an attempt to win back some of what they had lost in 1939. In 1942, fishermen struck over having to pay vessel insurance premiums, which were soaring due to wartime dangers. They also struck at General Seafoods Corporation the same year, demanding a closed shop. Between 1943 and 1944, fishermen also responded to broader national issues: they struck, threatened to strike, or periodically walked out in protest over price ceilings imposed on fish products by the federal Office of Price Administration. Nor did the end of World War II make much difference. In September 1945, owners refused to pay the crew of the F/V *Medford* standby pay for waiting to sail at a moment's notice. This sparked a general strike and owner-imposed lockouts across the fleet. Finally, between February and May 1946, fishermen struck once more, demanding a sixty-forty split of fishing trip profits instead of the then-current fifty-fifty split.[18] Unionization of the Boston trawler fleet's fishing crews signaled yet again that New England fishing was just as industrialized, modern, and mature as any other major American industry.

Mobilizing Modern Industrial Strength

By 1940, few engaged in New England's fisheries doubted that they were big business in Boston. Managers, financiers, fishermen, processors, wholesalers, and retailers all looked at the heavy trawler haddock fleet, responsible for the vast majority of New England groundfish landings, as a full-fledged industry. Simply put, the qualms that had wracked New Englanders in the 1920s and 1930s had eroded away. In their place came a comfortable, popular acceptance that Boston's trawler fleet was one of the world's most modern, efficient, and profitable fishing industries. Boston's fleet now enjoyed public support as it kicked itself free of the past.

The largest firms did not rest on their laurels, however. Beginning in the late 1930s, General Seafoods Corporation began planning for expansion, anticipating an increase in Midwestern demand for frozen fish. To meet this impending demand and to optimize profits, General Seafoods sought to expand production and lower production costs by bypassing Boston's fully unionized fishermen and setting up operations in Canada. This strategy followed the example of Gloucester's Gorton-Pew Company, which had operated a shore station on Canada's Magdalen Islands in the Gulf of St. Lawrence since the 1920s. With a cheaper, more docile labor force that was closer to healthier fish stocks, a Canadian plant would generate enough profit to offset the additional import duties for Canadian-caught fish from the Magdalen Islands. The firm also tried to leverage subsidies from the Canadian government, and in the late 1930s explored a similar deal with Newfoundland to set up a processing plant on the island's southern shore.[19]

The timing of these moves suggest that McHugh's successful leadership of the AFU provided significant motivation. General Seafoods representatives opened discussions with the Newfoundland government in 1937, just as Boston's fishermen turned out on strike. Newfoundland officials promised General Seafoods $200,000 in interest-free factory construction loans and people eager to work, given the island's high unemployment. In return, General Seafoods promised jobs for a region chronically short of them, along with a slug of American investment capital and construction contracts. The deal could only work, however, if labor savings and subsidies significantly offset added import costs. To gauge its potential costs, the firm made "oral requests" and informal inquiries to U.S. Treasury Department officers about whether fish products from an American-owned, Newfoundland-based-and-staffed plant could be imported into the United States duty-free. The Treasury Department referred to earlier cases involving the whaling industry to determine, in T.D. 49682—a statement quietly released in August 1938—that General Seafoods' Newfoundland products were "American fisheries products" and therefore not subject to import duties.[20]

The AFU belatedly got wind of this determination and immediately launched an investigation. Incensed that an American firm would export jobs to Canada, the AFU investigation concluded that T.D. 49682 reeked of political influence and corruption. The union pointed out that General Seafood's president, Clarence Francis, sat on the Business Advisory Council within the Department of Commerce, which also oversaw the BCF. Another General Seafood official, Gardner Poole, sat on the Fisheries Advisory Council of the BCF, a position that allowed him to "sell" one of General Seafoods' trawlers, F/V *Harvard*, to the bureau for one dollar in 1939 for scientific research purposes. The firm used these ties to avoid antitrust investigations when it purchased the Bay State Fishing Company outright for $1.2 million in October 1938 (and then promptly closed Bay State's Vinalhaven, Maine, processing plant). Most problematic, according to AFU findings, was the fact that General Seafoods' largest shareholder, Marjorie Davies, was married to Joseph E. Davies, who worked in the State Department. Between these connections and the quiet backroom agreement, the AFU determined that General Seafoods had used insider dealings and political connections to export American jobs in exchanged for shareholder profits.[21]

While there had been public debates in Newfoundland over the proposed deal, McHugh's AFU learned of these dealings only in early December 1939. Within two months, however, the union had secured support from the Massachusetts congressional delegation. The delegation cited lack of public transparency and called on the Treasury Department to rescind its findings. Treasury's refusal set in motion a congressional hearing on T.D. 49682 in March 1940. More than fifty witnesses united in opposition to the Treasury's decision. In April 1940, a bill sponsored by George Bates of Massachusetts appeared before the House Committee on Merchant Marine and Fisheries: it defined an "American Fishery" so as to prevent the General Seafoods deal from moving forward.[22]

Events surrounding General Seafoods' Newfoundland gambit, revealed in committee deliberations over Bates' bill, exposed important issues characterizing New England's fishing industry and public perceptions of it on the eve of World War II. Unlike earlier debates surrounding the fisheries, the 1940 hearings yielded few references to the heritage, tradition, or history of New England fishermen. Instead, most parties—including the fishermen—saw the industry in very modern terms. Not surprisingly, General Seafoods and Gorton-Pew based their arguments on profitability, creative interpretations of previous treaties, and modern precedent. Yet the congressional inquiry came about largely through the AFU's efforts, which marshaled its own political influence to preserve union authority, members' livelihoods, and the place of the industry in New England. Disdaining virtue, heroism, history and tradition, McHugh employed political and legal tools as sophisticated as those used by General Seafoods and Gorton-Pew.

The final decision on the bill represented a balance between powerful forces struggling in modern American industry. Because their operations were compatible with the fishing provisions of the 1818 treaty between the United States and Great Britain, and due to support from its processors' union, Gorton-Pew was allowed to maintain its shore stations on the Magdalen Islands as they had since the 1920s. By contrast, General Seafoods' argument that modern frozen fish processing should enjoy the same precedent found no support. In another major win for the AFU, General Seafoods was denied the right to bring its Newfoundland products into the United States tariff-free. In the 1940s, Boston fishing conglomerates and its unionized fishermen were equally willing to use political connections to advance their interests. Indeed, owners, fishermen, and the public almost expected them to do so.

Modern Fishing and New Critiques of the Past

New England had come to accept the modern fishing industry with its labor unrest and political machinations. But attitudes toward outmoded fishing centers suffered a dramatic turnabout in little more than a decade. Instead of celebrating the persistence of tradition, as in the 1920s, commentators outside the industry sharply criticized ports that failed to modernize by the early 1940s. Maxwell Frederic Coplan, writing about Gloucester for the *Saturday Evening Post,* broke from earlier historical sketches to critique heroic representations of the port. He questioned whether fishing could sustain the community in the future. Appearing in such a widely read publication, Coplan's critical review of Gloucester past and present fundamentally challenged how the broader public saw New England fishing.

Like earlier representations, Coplan rooted his discussion in the harbor's ancient, partly fabricated past. Citing Gloucester's dubious claim to hold the eleventh-century grave of a Viking, Thorwald Ericson, Coplan quickly moved to the harbor's earliest historical role in Samuel de Champlain's expeditions. Colonial fisheries, ruined by the revolution, revived until "the Civil War brought Gloucester its longest period of prosperous years." Here, however, Coplan deviated sharply from the port's established narrative. Rather than resulting from mythic Yankee commercial prowess, Gloucester's postbellum prosperity came because "the war and the wave of foreign immigration which followed gave the harbor what it so long awaited—a domestic market for its fish. And as its fishing fleet grew, provincial fishermen from Nova Scotia, Cape Breton, and Prince Edward Island filtered into town, and the Portuguese arrived to build their compact little colony."[23]

While Coplan claimed that "these were Gloucester's heroic years," he made clear that heroism came with a price: "It was a period of mounting tragedy, too, for, as the fleet increased, so did the casualties at sea. For years it was a rare month that did not take its toll of Gloucester fishermen." While fishermen's mortality often appeared in earlier histories and public commentary, never had the city's success been so closely tied to its human price. Indeed, Coplan wrote, "That was Gloucester, a town of a thousand personal tragedies, but one of great collective triumph over the sea."[24]

That triumph had come long before, however, when Gloucester annually landed over 60 million pounds (27,223 mt) of cod, haddock, and halibut, which sustained a "self-contained industry" of processors, outfitters, and wholesalers surrounding the harbor. By 1937, "Gloucester wharfed a little more than thirteen million pounds [5,900 mt] of cod, halibut, and haddock. In that figure lies the final tragedy of the town that waited so long and fought so hard for success." Reliance on salt fish, as opposed to fresh fish, and refusal to modernize sealed Gloucester's future. Boston's modern equipment, fresh fish markets, and gravitational pull on incoming landings left Gloucester as a second-rate fishing port. Only tourists, painters (often lured there to paint pictures of nostalgic decay), and Works Progress Administration–funded roads and bridges provided beacons of hope for the fading city.[25]

Coplan's article epitomized a regional trend that pushed New England's once-celebrated schooner fisheries firmly into the past. Modern fleets, processing plants, organized labor and management, and Washington lobbyists composed the new face of the industry; schooners and rotting piers no longer elicited sympathy as they had in the 1870s when J. W. Collins used such imagery to garner support for Gloucester's salt fish industry. By the 1930s, New England fishing interests sought for many reasons to present their industry and region in terms of cutting-edge industrial production and management. Gloucester's remnant fisheries were part of a fading past.

Expanding Fishing Pressure across the North Atlantic

Coplan may have been premature in writing Gloucester off, however. In 1933, landings of Acadian redfish (also known as rosefish or ocean perch) were not even recorded. Considered a trash fish, this small, spiny, long-lived, deep-water species appeared mostly as bycatch in the otter trawl fleet that had burgeoned over the previous two decades. That changed in 1934 when declining haddock landings compelled fishing captains to fill their holds with other species. Redfish landings rose accordingly from 300,000 pounds (136 mt) in 1933 to 1.3 million pounds (590 mt) the next year. Furthermore, processors found they could adapt their machinery to clean, package, and freeze redfish for Midwestern and southern markets, and fishermen began to land even more. In 1937, total New England catch rose to 17 million pounds (7,713 mt) and soared to 67 million pounds (30,399 mt) the following year. By 1940, landings reached 85 million pounds (38,566 mt). According to Rachel Carson, the noted environmentalist then serving as an aquatic biologist for the U.S. Fish and Wildlife Service, preliminary indications for the 1941 catch were significantly higher—almost 140 million pounds (63,520 mt).[26]

The redfish bonanza, for the speed of its origin and intensity of its expansion must be termed a bonanza, not only saved Gloucester as a fishing port; it also brought some prosperity to the medium and small draggers that had struggled through the 1930s. To illustrate, by 1937 Gloucester's midsized otter trawlers landed more redfish than any other staple species. Redfish landings beat out cod by roughly 5 million pounds (2,269 mt), mackerel by about 9 million pounds (4,083 mt), haddock by 1.5 million pounds (227 mt), and pollock, another high volume (if not highly valued) fishery, by 6 million pounds (2,722 mt).[27] On the eve of World War II, redfish had resurrected otter trawling and mechanical processing in Gloucester in just a few years, reversing three decades of decline.

New England fishermen outside of Boston and Gloucester also needed help to survive the Depression. During the heyday of the haddock fishery, the industry had created a new market for juvenile haddock, called scrod haddock. Trawlers soon fished immature haddock so intensely that William C. Herrington, in a 1941 report titled "A Crisis in the Haddock Fishery," cautioned the fleet against overfishing juvenile haddock.[28] Once again Herrington's calls went unheeded, and haddock stocks weakened through the 1940s as they had in the 1930s. Trapped between the Depression and falling haddock landings, outport New England fishermen turned to new species to sustain them.

This strategy of moving on to harvest new species as old stocks become depleted expanded human impact on local ecosystems at many levels. Flatfish, which live on the bottom with small mouths incapable of taking baited groundfishing hooks, had only been available in New England markets since the late nineteenth century. They were little known outside local fisheries before the advent of beam and otter trawling. Through the 1930s, however, they took on greater importance, offering abundant landings for vessels that needed to fish to survive. Flounders, as a proportion of New England's total catch across all gear and species, held steady around 4 percent between 1931 and 1938. During the same period, heavy trawlers targeting haddock on Georges Bank increased their flounder landings markedly. In those seven years, the large otter trawl fleet increased flounder catches from roughly 1 percent to just over 2 percent of the total New England fish catch. Small otter trawlers focused on flounders even more, expanding their portion of flounder landings from next to nothing in 1934 to 0.6 percent of total New England fish catch. Inshore draggers targeted the small flat fish so much in the 1930s that they came to be called "flounder boats" on the docks and in federal statistical reports.[29]

Comparing the changing proportion of a species' landings to the total catch brings a consideration of fishermen's aggregated choices into the analysis, independent of gear and of other landed species. This is because fishermen confront a series of opportunity costs when bringing fish to port. Does running in and landing one load of fish of a certain species, size, and value prevent fishermen from catching other, more valuable fish? Conversely, does it make sense to toss one load of fish overboard to make room for a potential catch of more valuable fish when staying out longer costs more money and may compromise the value of the fish still onboard? Both scenarios represent tradeoffs in terms of species composition, size, and value, and also time and cost.

Viewed this way, the total landings in a given year, across all species and gear types, reflect a compromise between what fishermen want to catch and what the ecosystem and their labor can provide. Landings also reflect what fishermen feel they *can* land. By the 1930s, buyers on the piers played increasingly important roles in dictating what species fishermen could land profitably and which ones they could not. Ultimately, the changing proportion of a species in aggregate landings shows not only fishermen's ability to find and land that kind of fish but also their changing perceptions of the market for each species, and how they engage the social organization of the industry.

Figure 1 compares some of the most important commercial species over the 1930s and reveals an important shift in how otter trawler captains viewed individual species' importance. For a number of reasons, large haddock, long the mainstay of heavy trawlers, slipped in proportion. Increased landings of market-sized cod, similar in size to large haddock, suggest that fishermen sought substitute species that could be processed by equipment designed to handle large haddock. Increased landings of both flounder and redfish reflect fishermen's desire to develop or take advantage of new markets.

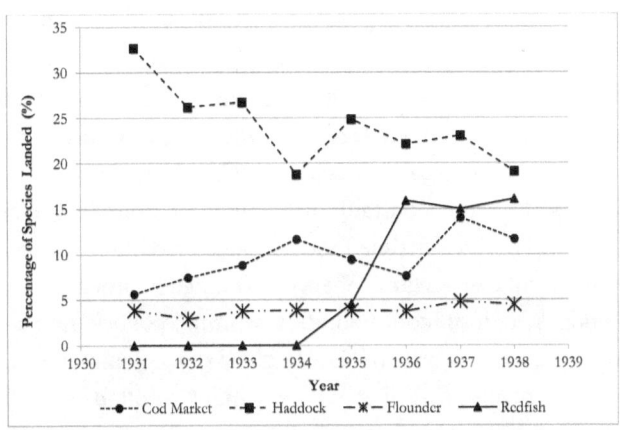

FIGURE 1. Proportion of select species within total New England fisheries landings, otter trawl fleet, 1931–1938. Source: Fiedler, "Fisheries of the New England States," 1931–38.

Shifting pressure onto new and different stocks also reflected the willingness of some ports and processors to take advantage of new opportunities, technologies, and markets. New Bedford is a prime example of how new species created new opportunities for fishing ports to develop. Unlike Boston and Gloucester, New Bedford shifted to fishing in the early 1930s by moving directly to otter trawls and working mostly on flounders and scallops. By the end of that decade, its fleet served thriving markets and profited handsomely from landing those species.[30]

World War II food rationing and production policies tossed new variables into the mix. Civilian beef consumption was restricted, and military contracts tasked New England fishermen to catch redfish to feed the troops. Americans contributed to the war effort by avoiding rationed meat and fish and instead consumed unrestricted fisheries products. Thus, wartime exigencies opened new military and civilian markets for fishing operations quick and nimble enough to adjust. By 1943, new markets and target species were keeping New England's fishing industry afloat. Fishing operations geared to supply large processors and wholesalers could not adapt as quickly as the small and medium dragger fleets. Fishing operations out of New Bedford and Point Judith, Rhode Island, featured smaller, more adaptable vessels, and their fisheries expanded markedly during the war.[31]

Governmental agencies actively encouraged people to consume underutilized species, and this campaign expanded fishing's impact across the marine food web. Writing for the U.S. Fish and Wildlife Service, Rachel Carson penned *Food from the Sea* in 1943 as a consumer's guide to selecting unusual fish products. Carson encouraged American "housewives" to serve nontraditional but more abundant species like redfish to help relieve wartime food shortages. Additionally, she showed readers that, while seven familiar species (ranked in order of size of landings: haddock, redfish, cod, flounders, whiting, pollock, and mackerel) comprised the vast majority of New England's 1940 commercial landings, sixteen other species contributed over a million pounds each, and fifty more supplied smaller landings, all of which made for good eating. These figures supported Carson's larger mission to convince American consumers that "scarcely any other class of food offers so great a variety—so rich an opportunity for gustatory adventure. The housewife who experiments with new fish species and new methods of preparation banishes mealtime monotony and provides delightful taste surprises for her family." Whether a "salad surprise" of "shrimp or flaked mackerel in gelatin" or a "fillet of wolfish" substituted for salmon steaks truly appealed to her wartime readers lies beyond this study. In advertising the availability of these species in the marketplace, along with cookbooks published by commercial fishery organizations as well as the Fish and Wildlife Service, Carson's work represented a conscious wartime effort to broaden the definitions of food fish. As government policies promoting the consumption of new fish species persisted well into the 1960s, human pressures across marine ecosystems reached new heights and remained high for decades to come.[32]

The Depression era strained fishermen to the breaking point as they struggled to ride out the economic crisis, but their survival strategies put even greater strain on local ecosystems. Otter trawling began to put tremendous pressure on haddock in the mid-1920s. After the economic collapse of the Depression, fishermen not only took more and younger haddock; they also targeted new species such as redfish, a variety of flounder species, and even cod. What landings data do not convey is how much of the fish was tossed overboard, mostly dead, as fishermen sought to optimize every day they spent fishing. Trawling and mechanical processing put New England's age-old fishery on a modern foundation, but industrial levels of raw materials—fish—had to be maintained to keep the rest of the processing operations running smoothly. Choosing which species the market wanted, and landing consistent supplies of those species to meet the demand, exacerbated the pressures on local marine ecosystems. Imbalance between economic demands and ecological realities proved to be the Achilles heel of New England's modern fishery.

New England's Two Groundfish Fisheries

Peace returned, lifting wartime rationing and price control policies and sending many heavy trawlers back out after haddock and now redfish. For small- and medium-sized vessel fleets, haddock and redfish remained important parts of their portfolio, but the newer fisheries developed during the war filled a unique, profitable niche. By the late 1940s, the two groundfish fisheries stood in stark relief to one another in many ways. Operating out of Boston and Gloucester, heavy trawlers worked on haddock and redfish stocks to feed high-volume processors producing frozen filets for nationwide distribution. Medium and small draggers still took advantage of the 1919 corruption settlement by landing their catch at Boston Fish Pier and adding their haddock and cod to the stream of industrial materials. In addition, the smaller draggers and flounder boats worked newer species not targeted by the heavy trawlers for local markets and small-scale processors. Draggers and flounder boats, offloading in the region's industrial fishing centers, found ready welcome in Maine and Cape Cod outports as these smaller boats' prosperity helped smaller ports reemerge in the 1930s and 1940s from decades of decline.

Other variables differentiated these fleets: capitalization levels; share structure; gear type; operational range; racial, ethnic, religious, and geographical divisions; level of unionization; level of organized owner associations; and—based largely on the two previous variables—political influence in Washington. This last facet was one of the most important. From World War II until the mid-1960s, owners, industry operatives, and union leaders from the heavy trawler fleet provided the most consistent voice for New England fishermen in federal policy hearings and discussions. Large and small draggers hailing from other ports, including Gloucester, rarely spoke in the debates of the 1950s and 1960s, preferring instead to follow the lead of the trained lobbyists and leaders working for the large concerns. Yet the power dynamic shifted as the heavy trawler fleet eventually declined and the dragger fleet remained viable because of its wider ecological foundation. The next two decades would make clear that political influence in New England fisheries rested on two critical foundations: an image acceptable to persistent cultural conventions about New England fisheries and the ecological, and therefore economic, strength of the stocks on which the fisheries relied. In other words, between 1946 and 1966, political power within New England fisheries rested on ecological sustainability and cultural suitability. The fleet that could don both mantles would have better success in shaping Washington's policy debates.

CHAPTER 5

THE CONSEQUENCES OF MODERN FISHING

Stock Declines, Labor Unrest, and International Competition

In early February 1947, the bottom fell out of the Boston fish market. As *Boston Globe* journalist Gerald Roscoe reported, "Fishermen accustomed to receiving as much as 23 cents a pound for haddock were faced with prices as rock-bottom as 3 and 4 cents a pound." To avoid cratering prices amid rumors that "'the golden era' of the fishing industry had come to an end," twelve Boston vessels refused to offload their catch and sailed for Gloucester *en masse* seeking a better return. They didn't get one.[1]

To some degree, the collapse was of the industry's own making. A fish-handlers' strike in New Bedford over weighing policies and oversight had recently closed that port to incoming vessels and sent several offshore redfish boats to Gloucester. In Gloucester, however, captains found shoreside processors already glutted with more redfish than they could handle. With little choice, the trawlers steamed for Boston seeking to sell their no longer quite so fresh catch. But other vessels had done the same thing. On most days, Boston Fish Pier could process the catch from thirty-five vessels. On that day in early February, however, forty-three vessels, "most carrying heavy loads," dumped their catches into the industrial refrigerators, swamping the market. Perhaps for the first time in the industry's history, efficient gear, large-scale investment, and organized workforces harvested more fish than even industrial processing facilities could handle. As prices crashed, fears rose, and some began to wonder whether New England's embrace of modern mechanized fishing had been such a good idea.[2]

Prices rose somewhat later that week, but the shock to the heavy trawler fleet was still palpable. Vessel owners and labor leaders knew that favorable wartime markets for fish, which had held through 1947, would soon face stiff international challenges, even without domestic landings glutting markets. The first manifestation of shifting markets appeared early that year. Prices for haddock, the backbone of New England's heavy trawler industry, declined steadily through January. Foreign frozen fish imports, mostly from Canadian producers expanding their U.S. markets, drove prices further downward. According to AFU president McHugh, local fishermen produced roughly 135 million pounds (61,252 mt) of fish in 1945, but the market imported another 43 million pounds (19,510 mt), roughly equaling one-third of domestic production, including from Canada. According to Thomas D. Rice, spokesman for the New England heavy trawler owners' association, Canadian producers sold imported cod for six to eight cents a pound cheaper than New England redfish fillets (labeled as "ocean perch"), cutting into sales in emerging Midwestern markets. In short, declining haddock prices, cheap Canadian cod imports, and overharvested New England redfish stocks had thoroughly swamped wholesale and retail markets by early 1947. Gloucester's freezing facilities warehoused 12 million pounds (5,445 mt) of redfish—six months of normal production capacity. Boston's freezers held 11 million pounds (4,991 mt) when normally they stocked only 7 to 8 million pounds (3,176–3,630 mt).[3]

Breaking the AFU

Within two weeks of February's market crash, collapsing prices resurrected the ghost of the 1919 Boston Fish Pier settlement, with its requirement that all fish be sold at auction, even if the buyer owned the vessel that caught them. The 1919 agreement also allowed captains and crews to sell their own catches where and if they chose to do so. Citing the latter provisions, and in the face of plummeting prices, twenty-five vessels in Boston's unionized heavy trawler fleet refused to offload catches totaling 2 million pounds (907 mt) until the processing firms agreed to pay a minimum price. On the surface, minimum prices would stabilize each vessel's bottom line on every trip, with increased profits returning to the owners, who were also the processers buying the fish in some cases. But processers made far more money retailing fish products than they made catching fish. In their calculus, minimum prices cut more deeply into wholesaling and retailing profits. As a result, Boston's large processors refused to pay, and AFU members refused to sell. The standoff quickly created a work stoppage that seized up New England's industrial fisheries for the next five months.[4]

An earlier, more bitter strike in 1946 had won McHugh's AFU membership a contract granting them a 25-percent greater piece of the profits at vessel owners' expense.[5] Now, with the bottom falling out of Boston's fish market, vessel owners enlisted the aid of Massachusetts attorney general Clarence A. Barnes to secure an injunction against the union's 1947 walkout. For vessel owners and fish buyers, the walkout created an opportunity to win back some of their previous year's losses. For Barnes, however, the walkout allowed him the means to undermine the AFU's power, which had increased significantly since the late 1930s. In a series of court cases, Barnes argued that the union's refusal to sell fish represented attempts by producers to fix market prices, ironically violating the same Clayton Anti-Trust Act that Boston processors had run afoul of in 1919.[6]

Barnes opened a series of investigations through the winter and spring of 1947. Their findings revealed the internal mechanisms the union used to withhold catches, intending either to even out fish supplies or to set prices, depending on one's perspective. As early as 1944, for example, the AFU had imposed catch limits as low as 2,000 pounds (0.9 mt) per person per vessel on large trawlers.[7] To enforce these rules, the union fined captains and crews who offloaded more than their imposed cap.[8] Barnes's investigation also found that in 1945, the union pressured nonunionized fishermen, in this case Italian boat fishermen, to limit their catches as well.[9] For the attorney general, AFU attempts to control the productive process had gone too far. According to the *Boston Globe*, "What the [AFU] have done is to use their control of virtually all the manpower engaged in the marketing of fresh fish in this Commonwealth (a control which is in itself lawful and harmless), and thereby, by arbitrary and unilateral action . . . create and maintain a monopoly whereby competition in the Commonwealth . . . may be restrained."[10]

New England's fishermen occupied an ambiguous place in modern labor relations. The legality of the previous AFU strikes and walkouts supported the interpretation that fishermen were industrial employees who enjoyed the right to collectively bargain. In terms of New England's traditional share-system, however, fishermen were independent producers, pooling labor and capital to mount a fishing voyage. Barnes's references to fishermen's independent producer status had nothing to do with tradition, however. Instead, he argued that the AFU's actions fell outside laws protecting unions' right to strike. Fishermen's refusal to sell their product posed a threat to the proper functioning of markets: everyone would eventually benefit if buyers had access to "unrestrained and unrestricted catches." McHugh also jettisoned tradition, responding that "this is an encounter between the union and the law of supply and demand: it is skirmish in the universal contest between the seller and the buyer, one of which is trying to sell as dearly and the other to buy as cheaply as he can."[11]

Ultimately, the union fell victim to the antitrust legislation, just as the Bay State Fishing Company had done thirty years before. In 1947, judges ruled against the AFU in a series of court cases that left McHugh's union intact but greatly hobbled. The courts determined that the AFU, while legal, represented producers, not employees. Therefore, refusal to offload catches and limiting catches through union mechanisms constituted anticompetitive practices. Fishermen could not legally restrict fish landings for economic or, as would be argued later, ecological concerns. Because fishermen, along with owners, were viewed as fellow producers in fishing trips, the court found that as many fish as possible had to be caught and landed as efficiently as possible, regardless of the consequences.[12]

The case against the AFU revealed the liabilities of Boston's modern heavy fishery. By consolidating wholesaling, processing, and vessel ownership in the 1920s, "buyers" had exercised significant control over Boston Fish Pier, where the vast majority of New England's fish came ashore. Fishermen had little choice but to organize and resist the commercial and political power arrayed against them. In stark contrast to New England's idealized fisherman from the 1920s—humble, hardworking, independent, but most importantly docile—New England's unionized trawler men soon flexed their collective muscles in what had become a chronic struggle between the market and the marine environment. Still, this evolution within an industry so long viewed as traditional appeared alien and unacceptable to many people. A *Time* magazine piece, titled "Labor: Monopoly Broken," crowed over "Home Run" Barnes's victory, using a nickname from his days playing ball at Yale as though the litigation had been a baseball game.[13] No more powerful than other industrial unions, McHugh's AFU had the temerity to move away from New England's fishing traditions, and the nation celebrated its loss.

The Contested Imageries of Changing Fisheries

The 1940s strikes alienated many observers who had come to celebrate New England's resurrected fishery over the past decade. For Milton McKaye, writing for the *Saturday Evening Post,* modernization had created more trouble than it was worth. Older methods—greater danger and lower pay notwithstanding—had proven far better for the region as a whole. In his piece "Good-By to the Dory Trotters," McKaye criticized in pointed detail the ways in which New England's heavy trawler fishery had evolved. "In 1945 surly New England fishermen, setting an all-time record, yanked more than 725,000,000 pounds [328,947 mt] of fish from the sea. In 1946 the catch was smaller because Boston, one of the great mother ports, was handcuffed for five months by a fishermen's strike." In fact, McKaye saw little worthy of praise in Boston's modern industry: "The industry has grown to a powerful and loose-jointed maturity during the last decade. Definitely big business in dollar terms, it continues to be as unruly as a Kilkenny picnic and as turbulent as Saturday night in a mining-town saloon."[14]

McKaye's problem with the modern fishing industry was its deviation from its humble past. High capital requirements had pushed fishing beyond the reach of a returning war veteran looking to use a GI loan "to set himself up in a tidy business." Instead, only large firms like General

Seafoods, Atlantic Coast Fisheries, Booth Fisheries, and Gorton-Pew could afford the modern, steel-hulled trawlers that could land enough fish to turn a profit. More problematic to McKaye was how modern labor organizations had shaped the fishing industry. Citing the recent strike-ending deal between the AFU and vessel owners, McKaye observed that "an ordinary deck hand on a Boston trawler can earn from $8000 to $10,000 a year. Even a friendly jury would be hard put to describe a deck hand's job as skilled." A captain made a lot more, from $18,000 to $30,000 yearly, "a fact calculated to induce heart-burn among our four-striper Navy captains."[15]

"The bugaboo of mechanization is always labor trouble," McKaye continued. While fishermen had made a hard and mean living in the past, McKaye contended that "the old, established operators in Boston and Gloucester have a record of fair dealing," though others "did not stack up so well." His fair-dealing old operators (McKaye ignored the 1919 indictments) deserved better than the unions had recently given them: "Over the last four and a half years, there have been fifty-eight strikes and work stoppages in Boston." Although McKaye lumped vessel owners' lockouts among the union's strikes, he estimated that the fishery lost roughly 30 million pounds (13,612 mt) of catch during the first three months of the 1946 strike, at a cost of $2.25 million to the community. Even among shoreside processors, organized labor was undermining industry viability: "The Sea Food Workers Union—AFL—is a separate entity from Pat McHugh's Atlantic Fishermen's Union, but it is tightly organized in New England and New York, and, with the turbulent labor scuffles of recent years in mind, many operators welcome mechanized filleting with distinct reservations."[16]

McKaye also felt that those currently working in the fishery gave it an alien air. Ironically critiquing romantic notions of New England's fishing present using romantic preconceptions of its past, McKaye lamented, "The lean and celebrated Yankee, tired of hardship, long ago yielded his place to newcomers, and New England fishing isn't what the romantics think it is. Today the Newfoundlanders—many of them of Irish stock—outnumber all other groups and maintain a dominant influence." Echoing earlier nativist assessments from the turn of the century, McKaye continued, "A good many of them hadn't even bothered with American citizenship until the irksome security regulations of World War II came along." Among these alleged interlopers McKaye added Italians, Portuguese, Nova Scotians, and "native-born" Americans, with "a few aging Scotsmen tossed in for salt and savor."[17]

McKaye's article reflected new uncertainties surrounding New England's modern fishery arising from industrial conflicts that marked the immediate postwar years. Returning to age-old fishing iconography, commentators such as McKaye resurrected conservative arguments from the early century that celebrated a docile labor force. They did so not only to challenge the inclusion of heavy trawling among New England's traditions but also to criticize fishermen seeking better pay and working conditions. The modern labor organization, mechanized fishing, and capitalization levels all highlighted how far from the romantic ideal New England fishing had strayed. Like Winfield Thompson had done in the 1890s, McKaye lamented the new reality of fishing.

Not everyone believed that modernization had gone too far, however. Gerald Roscoe's four-part feature in the *Boston Globe* celebrated New England's rise to the top among the nation's fisheries in terms of the industry's past and present. Running during the 1946 fishermen's strike, Roscoe's first installment detailed the otter trawl "revolution": how diesel engines allowed small boats to pull the new gear, and how filleting, quick freezing, and a new target—redfish—expanded markets in the Midwest. These industrial changes ended New England's traditional salt-fish fishery and paved the way for a "golden era of prosperity unprecedented in its three-century old existence." Later installments detailed how the modern fishery had changed New Englanders' attitudes toward the sea and the need for some regulation. Whereas "for New Englanders, the sea has become a veritable gold mine," fishermen were also "cognizant that the ocean's supply of fish is not inexhaustible, and that conservation, or 'wise usage,' may become necessary." Roscoe recounted how Boston became "one of the Greatest Fishing Ports in the World," and how Gloucester, once moribund and dependent on government aid, revived during the war. According to Roscoe, a U.S. Fish and Wildlife Service marketing specialist declared that "it has been like a Horatio Alger story, from rags to riches. Gloucester has had its ups and downs, but since 1938 it has been definitely up—way up."[18]

Others challenged the conservative backlash to McHugh's leadership, and a few even supported some of the union's gains. Writing in the *New Republic* in September 1947, James Higgins and Gordon Donald pointed out how fishermen's efforts to secure a better living drew the conservative ire of the region and its press. Invoking Nathaniel Hawthorne's *The Scarlet Letter*, Higgins and Donald wrote, "From editorial pulpits Boston newspapers preached on the scarlet sin of the AFU. Massachusetts' number-one labor-baiter, attorney general Clarence [Barnes], strode forth to do his duty." Duty, however, had little to do with Barnes's professed aversion to wasting natural wealth; rather, it merely served as a pretense for his crusade against organized labor: "The spoiled fish gave Barnes an occasion to act in the old New England name of moral purpose. Brandishing an antique state monopoly law, he smote the AFU."[19]

Higgins and Donald's satire reflected the fact that the 1947 dispute could be seen in different lights. The authors acknowledged the role that public opinion and popular sentiment played in shaping fisheries policies: "The fishermen's experience with the Boston newspapers has made them publicly minded. The success of their struggle depends on their ability to create public understanding of the curious organization of the oldest industry in American life."[20] While New England vessel owners began appealing to popular opinion in the press in the 1870s, the press coverage of fishermen's strikes and labor disputes in the 1930s and 1940s, often with quotes from McHugh himself, illustrated his willingness to embrace the press on behalf of fishing crews.

More importantly, Higgins and Donald's narrative challenged notions, persisting from the 1920s, that New England's traditional shares system and auction sales favored fishermen. The share system—or as they called it, the lay system—they argued, had enabled owners to oppress fishermen during the colonial period, often for generations: "[It] looked like a partnership—share and share alike between owners and fishermen. But it wasn't. Shrewd owners not only determined the size of the shares, but as the years concentrated control of the industry, owner-merchants also determined the selling price of the catch. They soon discovered the lay to be a sensitive device which could be manipulated easily to depress wages and increase profits." Tracing the history of New England's fishery to the present, the authors contended that life on fishing schooners lacked the rosy hue Rudyard Kipling had portrayed in *Captains Courageous*. Instead, "a lucky crew would be told that its share barely exceeded the price of the lousy groceries which the owner had put on board at the beginning of the voyage—and for which, in the tradition of the lay, the fishermen had to pay." Ultimately, the privations of the lay system drove New England men to seek other work, so that "by 1919 . . . there was scarcely a Yankee to be found among the crews. The phony lay system had driven them from the sea."[21]

Although this version of New England's fishing history departed radically from conventional narratives, it still retained elements long present in cultural caricatures of fisheries. Racist and nationalist overtones had marked popular representations of New England fishermen since the 1890s. Now they appeared in Higgins and Donald's story, as it equated changing ethnicities and nationalities with signs of decline: "[Yankee-held] places [aboard fishing vessels] were taken by Newfoundlanders and Nova Scotians. Italians, Portuguese and Icelanders followed. The desertion of the fishing banks by sea-faring Yankees had paralleled over the course of the years the slow decline of the fish industry as a whole."[22] In defending the AFU, Higgins and Donald used economic and industrial evidence paradoxically similar to that used by the union's enemies. While the traditional lay system had oppressed fishermen economically, modernization put power into the hands of ownership and marginalized fishermen in the marketplace. The amalgamation of industry, facilitated through modern organization, compelled the union to fight back. World War II finally brought fishermen a living wage. Although postwar competition and fish stock depletion posed challenges, the authors argued that the union would prevail in the future.

Yet the 1947 market shocks marked an important turning point in New England's fisheries development. Price collapse and labor unrest ended wartime prosperity. For industry outsiders, collapse and unrest also challenged the region's uneasy relationship with its fisheries. Modern fishing could not replicate on newer vessels the social norms imposed by New England's fishing past. Schooner fishermen's purported docile nature, their humility, and their deference to their social betters simply had not translated into the mechanized fishing industry. While Boston's prosperity in the 1930s and 1940s had masked growing tensions between labor and management, the occasional strike or lockout notwithstanding, the five-month strike in 1947 made clear to all that fishing suffered the same troubles as other American industries. New gear and new vessels, unions and formal owners' associations, changed the game economically and culturally. Many New Englanders, wondering where their traditional fisheries had gone, didn't like it.

The Ecological End of New England's Industrial Fishery

Although shoreside facilities were so awash in fish that prices had plummeted, some fishermen remained concerned that fishing intensity during the war had depleted local stocks. Such fears were nothing new: New England fishermen had long worried that the industry would overfish local stocks. Concerns about overfishing fueled opposition to new fishing gear throughout the nineteenth century, drove campaigns against southern New England weirs in the 1880s and 1890s, and formed a major tenet of the 1912 Gardner bill. Nevertheless, the chronic persistence of such fears suggests that they may have carried some truth. Evidence indicates that generations of fishermen customarily, if unwittingly, downsized their understandings of what constituted "healthy" commercial fish stocks.

By the mid- to late 1940s, observers outside the industry began taking such concerns more seriously as well. In December 1946, Edmund Gilligan reported in *Colliers Weekly* that "it's getting harder and harder to find cod and haddock in sufficient numbers. The reason is that the New England fishing grounds are practically fished out." Unlike earlier claims based on individual observations and assertions in public forums, new concerns about stock depletion rested on concrete evidence: "Once upon a time, the fleets took 350,000,000 pounds [158,802 mt] of marketable fish in a season on Georges Bank, the best of the old Yankee grounds. Now they have to go all the way to Sable Island, off Nova Scotia, to fill their pens. They fished on Banquereau, beyond Sable Island. By the end of this season, they will be going even farther." Gilligan blamed dragging with otter and Vigneron-Dahl trawls, both large and small, and worldwide hunger for the ever wider spread of local stock declines. Supporting Gilligan's claim was Bert Hemeon, the captain of a Gloucester dragger owned by Ben Pine and named *Columbia,* after Pine's racing schooner from the 1920s International Fisherman's Schooner Race. According to Hemeon, "Banquereau won't last much longer. And I don't know where we can go after that."[23]

New England schooners had long fished Canadian grounds in the salt cod, mackerel, and herring fisheries. Unlike New England's hook-and-line cod fishery, which managed to sustain itself for centuries, its otter trawl fleet faced a supply crisis after just twenty-five years of targeting haddock, redfish, and flounder species for industrial processing. Here was a major contradiction for the region's modern fishery. Certainly more efficient, more profitable, and safer, motorized vessels and gear had resurrected New England's fleets, both large and small. By 1949, however, cheap, diesel-powered vessels hauling ever-larger nets had also dramatically reduced local stocks. Even the inshore flounder fishery, pursued in earnest by small draggers only since the 1930s, witnessed dramatic declines in inshore abundance.[24] Facing a shortage of prey where they usually fished, New England fishermen ventured farther out after the quantities of fish needed to sustain high-volume industrial production. Once on Canadian grounds, New Englanders encountered Portuguese, French, and Spanish fishermen who had worked those grounds with equal intensity since the end of World War II.

Industry outsiders also feared New England's fish depletions for their economic implications. A study by Harvard economist Donald J. White made headlines in the *Boston Globe* on April 6, 1950, by detailing how stock declines had stifled the industrial growth of New England's fisheries. White argued that four threats faced the region's "oldest industry" in the postwar market: scarcity of fish, tension between labor and management, marketing difficulties in the Midwest, and cheaper Canadian imports. Of those four, White identified local scarcity of commercially important fish as the most ominous. Scarcity not only threatened the flow of raw materials to processing plants; it also significantly raised productions costs. Furthermore, while scientific assessments were underway, "much of the evidence uncovered to date indicates that overfishing is an important factor in [fish stocks'] declining numbers."[25] White was not concerned with cod stocks. With the passing of the salt fishery in the 1930s, cod had slipped in economic importance. In the early 1950s, haddock and redfish remained the most valuable stocks, as they had been before World War II.

By the late 1940s, New Bedford's relatively young but significant fishery for yellowtail flounder also showed signs of distress. In addition to his concerns over haddock and redfish stocks, White felt compelled to identify yellowtail flounder as overfished and imperiled.[26]

To be clear, these concerns stemmed from the economic consequences of fish scarcity for New England's industrial fisheries, not necessarily from a latent environmental consciousness. And there was good reason for such concern. According to White's analyses, continued prosecution of the scrod haddock fishery had driven depletion of local haddock stocks—the same fishery William Herrington had cautioned against in 1941. After nearly a decade of ignoring Herrington's cautions, White concluded that "the depletion of baby haddock on Georges Bank through heavy scrod fishing has reduced the haddock population on that fishing ground by 62 percent since World War I." Those declines in turn sent the industry into Canadian waters, where fishermen continued to target scrod haddock. As early as 1934, White wrote, one Fish and Wildlife official had estimated that longer trips to the Nova Scotian banks added more than 2,600 fishing days to the Boston fleet's annual sea time. At roughly $250 per day, depletion of local stocks added $650,000 per year to the industry's overhead costs. Fifteen years and a world war later, those costs had increased significantly, and White concluded, "Since the present costs are two to three times greater than those in 1934 and Nova Scotia trips are even more common, Boston owners and crews today face a substantially greater disadvantage" from higher production costs than those experienced by other producers.[27]

Redfish and yellowtail flounder fisheries experienced similar problems. By 1941, only seven years after its first commercialization, the Gulf of Maine redfish fishery began to see declining catches despite increased effort. After that, the fleet had only been able to maintain supplies with Canadian fares. White cited other problems confronting the fishery: "The critical factor in the dismal outlook for redfish is their slow rate of growth." Since the fish took roughly nine years to reach sexual maturity, redfish populations required much more time to recover. The outlook of the New England small-boat yellowtail flounder fishery appeared equally grim. Targeted in earnest only since New Bedford became a major fishing port in the 1930s, yellowtail flounder landings had peaked in 1942 at 36.7 million pounds (16,651 mt) and then declined. Fishermen maintained landings by steaming farther out onto Georges Bank and by fishing southern New England grounds more intensely. Still, daily catches for small draggers working out of New Bedford declined from 9,600 pounds (4.4 mt) per day in 1942 to half that amount, 4,800 pounds (2.2 mt) per day, in 1948.[28] Never again could New England fishermen land the quantities of flatfish from local grounds that were commonplace in the 1930s. In less than a decade and a half, New England's otter trawl fisheries had pushed the ecosystem beyond what it could consistently produce.

Romanticization and Internationalization of New England's Fisheries

By 1948, concerns over the health of northwest Atlantic fisheries prompted the United States to host a November meeting of nations whose fisheries worked those grounds to see if international controls could be enacted. These discussions, among the first to address declining stocks of haddock, yellowtail flounder, and other commercially valuable species, once more brought New England's fisheries politics onto an international stage. For vessel owners and union leaders, the 1948 International Convention on Northwest Atlantic Fisheries (ICNAF) raised the political stakes of fisheries beyond regional concerns. The treaty also brought fisheries leaders into closer contact with State Department officials. If recent years had witnessed fishermen wielding power beyond acceptable limits, these meetings united labor and management in a common cause. Joining forces in the face of more direct federal oversight, labor and management rallied behind images of tradition to shape international treaties to their own ends. ICNAF and the domestic legislation that implemented it would not only reverse decades of labor unrest; they would also help resurrect and entrench New England's well-worn fishing traditions. Those myths would now stand at the heart of U.S. international fisheries negotiations.[29]

All treaty participants accepted that after only two decades of heavy trawler fishing, high seas stocks of haddock and other fish had begun to show signs of depletion. With that understanding, nations fishing in the northwest Atlantic shared a desire to sustain these fisheries both biologically and economically. In New England, prospects for meaningful State Department influence over regional fisheries practices provided a good and clear rationale for labor and management to set aside recent acrimony. Vessel owners and the AFU recognized the need to come together to support and shape international agreements to prevent overfishing in the northwest Atlantic.[30] Ultimately, the treaty brought together state-level regulators and federal departments and agencies as ICNAF participants worked through international agreements that aimed to reverse declining stocks.

In federal hearings, New England's vessel owners' representatives and unionized fishermen openly supported the treaty—an unusual moment of accord after a decade of near-constant strife. Speaking on behalf of the Federated Fishing Vessels of New England and New York, which represented the large firms owning Boston's offshore trawlers, Thomas D. Rice stated, "The matter [of marine fisheries depletion] has been coming to a head for a long time and something has to be done." Similarly, Patrick McHugh highlighted his union's long-term concerns over stock decline: "We've been advocating some sort of international control [over offshore fishing] for ten years. . . . Perhaps this meeting will come up with the right answer."[31] Additionally, representatives from other trade organizations and labor unions, along with representatives from coastal states, united in support of the treaty's final draft.[32]

Their positions changed, however, once the treaty had been ratified. As implementing legislation moved forward through Congress, both labor and management took issue with how the treaty's terms would blend with U.S. law and regulations. Most importantly, management and labor balked in unison about what preserving the fisheries actually meant. In the treaty and its implementing legislation, its sponsoring senator, Rhode Island's Theodore F. Green, was ambiguous about whether preserving "fisheries" meant the biological preservation of stocks or the economic preservation of the industry. Problematic for fisheries leaders, Green's language suggested that the term might encompass both meanings. Addressing the opening of the hearings, Green stated, "Work under the treaty should result in stopping the decline in the New England bank fisheries, enabling them to produce their maximum sustained harvest in the future."[33]

For others, however, preserving the fisheries was a strictly biological mandate that would eventually benefit the industry. Donald White, for example, whose research Green praised and included in the treaty's public record, first defined the concept in biological terms and then in industrial terms. He concluded that "the threatened depletion of New England's fishing grounds requires immediate attention."[34] So, too, did W. M. Chapman, special assistant for fisheries and wildlife to the undersecretary of state: "Depletion, however, has been noted, especially in the banks closest to New England where the fishing-out of the banks is said to have become acute." Further worrying to both vessel owners and labor leaders was the fact that according to State Department interpretation, neither the treaty nor its implementing legislation carried any economic authority at all. As Chapman further testified, "The treaty is a fisheries conservation treaty, and the bill a fisheries conservation bill. They are not intended to and do not in any way empower [U.S.] Commissioners to institute production controls and economic regulation or intrude into the field of labor-management relations.... They are for the purpose of producing the maximum poundage of fish out of the sea."[35]

Even Chapman's ostensibly clear separation of conservation from industry regulation proved to be muddied. That ambiguity made allies of McHugh and Rice. For industry leaders, maximizing seafood production represented biological and industrial issues that the treaty and its implementing legislation would necessarily affect. Several industry leaders accepted the need for biological conservation measures for fish stocks but defined "preserving the fisheries" in economic as well as biological terms. For them, preserving industry profitability was as important a "conservation" concept as preserving stocks of fish. The Fulham Brothers, Neptune Trawling, and Triton Trawling companies issued a joint statement demonstrating the dual meanings that the treaty's language carried in New England: "The immediate operation of the Northwest Atlantic Fisheries Convention [is] vitally needed to protect our natural resources and preserve the economic life of our industry." L. T. Hopkinson, president of the Atlantic Coast Fisheries Company of Boston, rejected the treaty's conservation implications but supported the need for international research into depletion: "We feel that the banks on which our vessels operate should be thoroughly investigated. The information at hand is not sufficient to support or refute the various claims for or against the need of restrictions on fishing."[36]

Others saw the treaty's value solely in the economic preservation of the heavy trawler fleet. State officials, technically distinct from industry, joined industry in this position. Francis W. Sargent, director of Massachusetts's Division of Marine Fisheries, stated, "By means of this convention, international controls can be effectuated that will be of far-reaching importance to the large fishing fleets of Massachusetts that constitute more than 80 percent of the offshore North Atlantic United States fishery.... [The treaty's implementation will] render a comprehensive service to the future of those who gain their livelihood directly and indirectly from the sea."[37] McHugh and Rice shared this interpretation. Neither was interested in seeing biological conservation limit fishing operations on the high seas. To the extent that biological conservation and industrial preservation could be separated, McHugh and Rice were about industry preservation alone.

Ultimately, however, the trawler industry obtained little from the hearings. The congressional bill clarified three important points: (1) the State Department, not state officials or local fisheries representatives, would interpret the language of the treaty and the legislation; (2) the State Department would define conservation in biological terms; and (3) the U.S. delegation would be dominated by state and federal officials likely to feel the same way. And as the treaty stipulated that national delegations would be composed of three members, federal officials stipulated that the U.S. delegation would include a State Department official, a representative from a state bordering the convention area, and a representative from the public at large. The delegation would not be dominated by industry leaders or lobbyists. Even with that broad composition, however, the dominance of government officers made it likely that the U.S. delegation would continue to define conservation solely in biological terms. Vessel owners and labor leaders alike now feared that their initial support of the treaty might have been misplaced. It seemed inevitable that ICNAF would regulate and restrict high-seas fleets.[38]

Right away, Rice and McHugh challenged the structure of the U.S. delegation, as well as the premise that implementing the treaty authorized offshore regulations. Rice made quite clear the opposition of vessel owners: "I have been instructed ... to appear before this committee and have recorded in the minutes of the committee hearing the unqualified opposition of the Federated Fishing Boats [sic] of New England and New York, Inc., to any international treaty [concerning northwest Atlantic fisheries]." Rice's employers felt that the treaty ignored the effect possible regional closures could have on local markets. Operating on the most heavily fished, or even overfished, areas within the treaty, Boston fleet owners feared that conservation measures would call for closing grounds close to home, forcing offshore trawlers onto more distant grounds. Closures near New England, while beneficial for conserving marine fish stocks, would also favor Canadian competitors, who were closer to healthier stocks and could undersell domestic markets.[39] Once lost to Canadian imports, Boston's fish markets could no longer sustain New England's more expensive fish products.

McHugh also objected to the treaty but targeted his criticisms to the structure of the delegation. Opening his comments with the AFU's 1938 founding commitment to the conservation of marine resources, he reminded the committee that the union had worked with William C. Herrington and regional congressmen to establish some federal regulatory authority over the offshore fisheries. He did not, however, mention the industry's refusal to abide by Herrington's recommendations for larger mesh sizes—the consequences of which were now coming home to roost. Instead, McHugh highlighted how the union had long introduced and shepherded resolutions for natural resource management in annual American Federation of Labor conventions.[40]

Green's drafting of the ICNAF's implementing legislation failed to provide McHugh the solution he desired. It was not the regulation of the high seas that alienated the union but that the regulations would emerge from shared jurisdiction: "During all this time we had one thought in mind that the Federal Government and it alone was to be the regulatory power and at no time did we contemplate having any [individual] State participation." To fill the U.S. delegation called for by the treaty, Green's bill tasked state fisheries commissioners with nominating one of the three members and the State Department for the second. That left only one possible seat to be filled by a delegate from the industry, which to McHugh reflected an unacceptable marginalization of fishermen's voices. Furthermore, to McHugh, the dominance of political appointees to the U.S. delegation represented a radical departure from tradition and, more specifically, the constitutional checks and balances on state power: "This bill is the beginning of an attempt to establish new and novel State intervention in the high-sea commercial fisheries of New England which are at the same time interstate and foreign commerce." Consequently, McHugh contended, "The fishing industry . . . [has] no desire to be guinea pigs for any form of academic experimentalism [in international fisheries management].... We do not believe that the lives and economic destiny of both fishermen and vessel owners should be the toy of a group of people who like to play with political theory."[41]

To counter the State Department's domination of the U.S. delegation, McHugh resurrected discussions of New England's fishing traditions that, up to this point, had been entirely absent from discussions of the treaty and its implementing legislation. To oppose the establishment of this new regulatory regime, McHugh argued that these regulations violated New England fishing tradition: "This [bill] means the complete regulation of our economic life and status—a method of life thoroughly foreign to the entire tradition of the New England industry—and we feel that it is entirely too much power to be entrusted to three people who have not any direct interest in the industry." Nor was he alone in his suspicion of the interest of state officials in regulation. New Bedford vessel owner and deep-water captain Randolph Matland testified, albeit erroneously given decades of state and federal investment in fisheries research, that "the state of Massachusetts and other States have never done the slightest thing for our welfare, and in fact the only thing they have ever done is 'hamstring' us. . . . We want no State interference without our affairs beyond the 3-mile limit [of state waters]." Well aware of the limitations of previous federal attempts at high-seas fisheries regulations, Matland continued, "If we must be regulated, let the Federal Government do it as they have always done in the past."[42] If regulation was needed, all components of industrial fisheries seemed to agree, the federal government alone should do it, and it should be done in accordance with New England's traditions. Given the recent history of high-seas fishing in New England, those traditions meant little or no regulation at all.

With little faith in individual state representation, and with regulation seen as an alien intrusion on New England fishing tradition, it was hardly surprising that industry representatives preferred self-regulation, if regulations were indeed inevitable. To support their argument, industry leaders pointed to the implementing legislation itself. In addition to providing one seat at large on the formal U.S. treaty delegation, the bill created an advisory body broadly constituted from among constituents who were directly affected by the regulatory agreements enacted under treaty provisions.

Such provisions were not novel to the ICNAF legislation. Since the 1920s, several West Coast fisheries had been managed by international agreement between the United States and Canada, which had included such an arrangement. Throughout the hearings, Chapman and other State Department representatives highlighted how well this advisory structure had served U.S. industry interests. New England fishing interests, however, objected to serving only in advisory and minority voting roles. Because the states had never before shown an interest in managing offshore fishing beyond the three-mile limit, industry leaders argued, those states did not deserve special representation on the U.S. delegation. McHugh, in particular, sought provisions that would grant fishing industry representatives unique and unprecedented authority over how the nation would meet its international treaty obligations under ICNAF. Offering little to justify his position, McHugh presented a romantic notion that only those who worked the sea had the knowledge needed to manage marine activities: "This new venture in conservation practice is too risky and too serious to be hamstrung by people who know little or nothing about practical fishing operation or the problems of the industry as a whole. [The state representatives] do not possess the experience either of vessel owners or of fishermen."[43] The fact that the treaty conferred no authority to regulate industry operations, McHugh felt, meant that fishermen should continue to regulate themselves.

This led to his objection to the treaty's operation. If only fishermen and vessel owners had the practical experience needed to manage the industry, it seemed logical to McHugh that they should dominate the U.S. treaty delegation tasked with negotiating conservation measures. In an audacious move, McHugh proposed that instead of State Department, adjacent state, and general public representation, labor and management should have two of the three delegation seats. In effect, McHugh's proposal would have granted private entities the power to effectively block government participation within an international treaty, neutralizing the government's position on votes cast by the delegation. This was nothing short of a private industry's usurpation of authority over U.S. foreign policy, if only in fisheries. McHugh justified his gambit by citing the State Department's desire for industry to support the treaty: "We have no confidence now and will not have any if this bill is jammed through [in its current form] as apparently there is the intention to do." If the State Department wanted industry support, they must grant industry special representation in foreign-policy matters. "The simple way to restore confidence," he argued, "is enlistment of vessel owners and fishermen in fundamental decisions both as to policy and practical administration of the convention."[44]

In the end, ICNAF was enacted without addressing most of the industry's demands. Federal and state officers still dominated the U.S. delegation. One public citizen had a seat, and industry representatives with no formal voting power spoke through an advisory body. Nonetheless, even without votes, fisheries representatives on the advisory panel could attend each meeting and influence the discussion. This significant concession brought New England's fishing interests deep into the center of regulatory power and gave them more influence than they had previously enjoyed. Equally significant, however, was how New England's fishing unions willingly invoked romantic images of fishermen's historic independence in making their case against the ICNAF framework. Even though his union symbolized fishing's modern development, McHugh's turn to the tropes of the past demonstrated the staying power of those images. Modern or not, accepted or not, New England's fisheries nostalgia still made for good politics. By the early 1950s, previous divides between the modern and traditional were beginning to blend, significantly influencing both international and domestic fisheries deliberations.

CHAPTER 6

INVOKING THE PAST, IGNORING THE PRESENT, COMPROMISING THE FUTURE

As the 1950s unfolded, New England's heavy trawler fleet once more abandoned its periodic attempts to fit in with and be seen like other American industries. Much as they had done in the 1930s and 1940s, industrialized fisheries embraced romantic imagery to gain national political leverage as they tried to reverse international agreements and tariff policies that disadvantaged New England fishermen in favor of America's Cold War allies. Both Thomas Rice and Patrick McHugh (and his successors) questioned, during tense and difficult debates, the fundamental relationship between New England fishing industries and state and federal government, particularly the legitimacy of any government oversight at all of offshore fisheries. Antiregulatory attitudes had been widespread in the fishing industry since at least the 1870s. Federal research and construction subsidies notwithstanding, New England's fishing representatives repeated one refrain throughout the 1950s—that federal authorities either restrained or ignored the fisheries.

Rice and McHugh turned to the Merchant Marine Act of 1920 (also known as the Jones Act), which had legitimized romantic notions of the unique maritime experience to thwart ICNAF.[1] Regardless of context and issue, the 1920 act declared that experienced mariners were uniquely and solely qualified to assess maritime matters. This legal tenet enabled Rice and McHugh to argue that fishermen were the only ones qualified to manage fisheries. Combined with the premise that government customarily spurns fishermen, it was easy for fisheries leaders to claim that regulation was alien to New England fishing tradition, therefore only fishermen had the practical expertise to judge fisheries matters. Should the regulation of offshore fisheries become necessary, fishermen deserved special consideration by virtue of experience and legal precedent in the conduct of U.S. domestic or foreign policy.

By using romantic arguments from the past to shape contentious issues in the present, McHugh and Rice developed a powerful rhetorical and political strategy to fisheries management debates. As cultural representations and the local press became even more sympathetic to fishermen's claims in the late 1950s, McHugh's and Rice's combination of romanticism and rhetoric gained power. Even though New England's haddock industry had grown in size, capacity, and complexity to become the most extractive fishery the region had ever seen, McHugh and Rice easily enlisted new voices, particularly from Gloucester, to give their image of traditional and artisanal fishing more credibility. Initially, their campaign yielded mixed results. But as Gloucester became more involved, and its adherents mobilized and deployed their romantic imagery, federal debates began to shift toward industry positions. Certainly, the ready invocation of hackneyed New England tropes was not the only reason industry saw gains by the end of the decade. Still, those tropes had acquired cultural legitimacy, if not authenticity, with grassroots constituents. At the end of the decade, elected officials found that increasingly vehement arguments based on "tradition" were increasingly difficult to ignore. This fertile combination of heritage, combativeness, and persistence emboldened New England's heavy trawler fleet to challenge federal authority over foreign policy. The 1950s debates revealed how McHugh and Rice joined with Gloucester's fishing leaders to develop a powerful and compelling argument that would shape New England public perceptions of the industry and industry lobbying for years to come.

Ecology and Unrest in Postwar Fisheries

Despite international accords and heightened public and industry awareness, fears about the health of New England's groundfish fishery persisted. Marked declines in haddock catches, offset somewhat by the targeting of juvenile haddock in New England waters, gave government researchers good reason to continue investigating the haddock fishery's sustainability. According to U.S. Fish and Wildlife biologist Ernest Premetz, between 1947 and 1951, otter trawlers working on Georges Bank and the Great South Channel destroyed roughly 4.5 million pounds (2,042 mt) of undersized haddock each year, about 6 million individual fish, most of which were less than a year old.[2] Before Premetz's analyses were completed, however, another Fish and Wildlife biologist working on the New England haddock fisheries, Howard A. Schuck, published cautionary articles in industry publications urging the fishery to curb such waste. Entitled "Protecting Baby Scrod Raises Production," "Destruction of Baby Haddock on Georges Bank," and "Current Haddock Situation on Georges Bank," Schuck's pieces sought support from various segments of the groundfish industry for the mesh restrictions and minimum harvestable fish sizes that ICNAF was considering at the time.[3] Like William C. Herrington's prewar recommendations, postwar attempts by Donald J. White, Premetz, Schuck, and others to make haddock fishing less destructive were "spurned" by the industry, as White himself put it, and failed to change its behavior.[4] New England's heavy trawlers continued to work on juvenile haddock stocks with small-mesh nets to maintain production levels. Despite gloomy prognoses, groundfish landings held steady throughout the 1950s. But as future analyses would show, fishermen maintained these production levels through increasing effort, targeting smaller fish, and bearing higher overhead costs. As in the 1930s, New England's fishermen were once again driving down local stocks of groundfish that were critical to their economic survival.

Landings data masked dire trends in the haddock fishery, but the fate of the redfish fishery was quite clear. After only fifteen years of exploitation, redfish stocks had crashed, and the resulting ecological changes pervaded both human and marine communities. Like haddock, collapsing redfish stocks led to controversy, conflict, and distress in cities and towns deeply tied to the resource.[5] On May 1, 1950, for example, 1,400 processing plant workers, members of Gloucester's Seafood Workers Union under the American Federation of Labor, turned out on strike to call for stabilized compensation and an end to piecework. In the past, fishermen's unions had from time to time found common cause with processing plant workers and turned out in sympathy strikes. Labor tensions in the fishing industry had been rare in Gloucester since the port revived in the mid-1930s, however, largely because healthy redfish stocks had meant prosperity for processing plants and local fishing operations large and small. Whereas in Boston unions battled management almost annually, in Gloucester, AFU management of the auction house, the predominance of vessel owner-operators, and more constructive relationships between the seafood union and the processors mitigated problems that might have led to confrontation.[6]

Relative peace, however, depended on the health of local resources. Once redfish stocks plummeted, vessels had to burn more fuel to fill their holds. The new prosperity faded, and fishermen and owners looked to each other for blame and redress. Fishermen were already trying to catch more fish. Processing plant managers, needing to maintain supplies and fulfill wholesale contracts, had little room to negotiate. Ecological strain induced industrial strain in short order, and only the nonunion shops, including Gorton-Pew, Davis Brothers, the O'Hara fish processing firms, and two others, remained working in the spring of 1950.[7]

As the processors' strike in Gloucester wore on into the summer, the future of Gloucester's fisheries once more seemed in doubt. Mayor John J. Burke, himself a vessel owner, ship chandler, and fish broker, sided with the strikers and provided paid municipal work for out-of-work fish processors. Others, however, questioned whether depleted stocks of haddock and redfish could still justify investment in large-scale processing plants. Moreover, the *Boston Globe* reported that some observers wondered whether redfish and haddock stocks could sustain the heavy trawler fishery at all. Some businesses considered giving up altogether, and a few firms affected by the strike discussed moving out of Gloucester and relocating operations closer to healthier redfish stocks.[8]

Once Gloucester's strike exposed the link between stock health and the future of the processing plants, other tightly focused strikes occurred in the region's major fishing ports. Simply put, industrial uncertainty compounded the ecological uncertainty surrounding New England's fisheries.[9] In reality, however, the two forms of risk were one and the same. Labor unrest in the 1950s fisheries stemmed from the economic consequences of overfished stocks because depleted local stocks undermined New England's competitiveness in a globalizing marketplace. As vessels had to increase production costs by venturing farther away to fill their holds, foreign producers situated closer to healthier fishing grounds could operate at lower costs. Consequently, foreigners could undersell American fish producers in American markets.

Local stock depletions in New England and Cold War politics both favored foreign producers, who took full advantage of those opportunities. Canada had undersold New England fish products in American markets since the late nineteenth century, even when levied with protective tariffs. Adding to that pressure, as part of the Marshall Plan following World War II, President Dwight D. Eisenhower transferred American fishing technologies and opened American markets to European allies in order to bolster resistance to the Soviet Union. These policies lowered prewar tariffs on imported frozen fish fillets, making former allies into adversaries against whom New England fishermen had to compete.

The Fish Stick Revolution and the Cascading Consequences of Ecological Decline

In 1953, a whole new food gimmick, the frozen fish stick, appeared on American plates. Fisk sticks were made from boneless, skinless pieces of white fish (cod, haddock, hake, and pollock, among other species) that were pressed into hundred-pound blocks and frozen solid. Using industrial food-grade band saws, the blocks were cut into perfectly consistent shapes, breaded, fried, frozen again, packaged, and shipped. For consumers, fish sticks made cooking fish—once messy, smelly and difficult—clean, convenient, and hassle-free. For producers, fish sticks offered a means to add lesser-utilized species to the industrial production stream. In short, fish sticks were an industrial product made by compressing diverse species, of different sizes, from different ecosystems, into the indistinguishable blocks from they were cut. The resulting uniform product emerged to win wide public appeal. Within a few years, fish sticks led to a surge in American fish consumption.[10]

New England fishermen missed out on that revolution, however. Local fish processors readily adapted their operations to process frozen fish blocks instead of fillets. But local fishing firms, locked into industrial facilities from the 1930s, were reluctant to retool to produce frozen fish blocks. Given the low overhead costs in Canada and Europe, few New England firms could have done so and remained competitive. Fish blocks even changed the politics of New England fisheries, separating New England fishermen from their erstwhile fish processing allies. Since fish blocks were neither frozen individual filets nor fresh-caught fish, they fell outside standing tariff structures. Consequently, New England processors happily imported cheap bricks of foreign mystery fish. Although fishermen sought to expand tariffs to cover fish blocks, U.S. processing firms, geared to fish stick production, were eager to maintain the status quo. With local haddock and redfish stocks collapsing after fifteen years of heavy fishing, the fish block "revolution" was the last thing New England fishermen needed.[11]

The combined effects of fish blocks and labor unrest became apparent in the early 1960s, as federal agencies continued their chronic struggle to cure New England's fishing woes. In analyzing the performance of fisheries from 1947 to 1958, federal economists found that stock depletion threatened to implode the heavy trawler fleet: "This postwar crisis has been marked by both diminishing catch and a price structure that has been inadequate to compensate for the lower domestic supply and the higher costs of vessel operations."[12] Due to declining local stocks, increasing foreign competition, and higher overhead, New England's large firms hemorrhaged money. Values of landed fish plummeted $10 million from 1948 to 1957. Lower redfish revenues accounted for almost half the loss, with the cod and haddock fisheries responsible for most of the remainder. Gorton-Pew, for example, lost $237,000 in 1952 alone. Other processors turned to the cheaper imported fish blocks by 1954 to offset the higher costs of domestic raw materials. More ominous for New England's heavy groundfish industry, by 1956 Gorton-Pew was operating three fish plants in Canada to provide frozen fish blocks to their American processing plants.[13] Production overseas was so much cheaper—in terms of labor, overhead, and industrial operations—there was no rationale for maintaining U.S. facilities.

The economic costs of overfishing, intense foreign competition, and labor unrest convinced many heavy trawler owners that it was time to get out if they still could. In a pattern recalling the consolidation of Gloucester's schooner fleet in the early 1900s, firms relocated operations or folded outright. Vessel owners cited labor problems when they sold three trawlers out of the Boston fleet in April 1949. This was only the beginning. Full on catastrophe broke out in 1953, when General Seafoods ceased its Boston operations altogether and sold off its twenty-six vessels.[14] After peaking in the 1930s with over one hundred vessels, by 1957 Boston's large trawler fleet numbered only fifty.[15]

Some firms moved their operations to be closer to healthier stocks; abundant fish and lower overhead more than made up for increased import duties. Nor was this unprecedented: as the 1940 debates revealed, Gorton-Pew had shore processing stations for salt fish on Canada's Magdalen Islands in the 1920s. In addition, Boston's five-month-long strike in 1946 drove many processors to reconsider the location of their operations. According to Thomas Fulham, who spoke on behalf of a large trawling firm, "The ability to process fish was transferred [from New England] to Nova Scotia and Newfoundland during the 21 weeks of the strike." Vessel owners also leapt at chances to get out of New England. Fulham continued, "Practically our entire fleet of fishing trawlers was offered to Uncle Sam to buy," when during the Cold War the government wanted to purchase vessels for its European allies. Eventually, twelve ships were bought for use in Germany, and the owners would have welcomed other opportunities to sell out as well: "We all, even in 1947, wanted to get out of the business, because the fleet owners could see the writing on the wall."[16]

As the heavy trawler fleet imploded, fishermen and vessel owners looked to each other for blame. Fishing representatives, focusing on the industrial symptoms of decline rather than its ecological causes, blamed union contracts and power. Labor leaders thought owners demanded too much from the crews. Yet both sides placed most of the blame on cheaper foreign imports benefiting from healthier local fish stocks and lower labor rates. In 1951, the Massachusetts Federation of Labor passed a resolution "to prevent transfers of certain operations to Canada for the purpose of taking advantage of wage levels 50–75 percent lower than those prevailing in New England."[17] Furthermore, labor and vessel owners united to lobby the U.S. Tariff Commission for higher tariffs for most forms of imported fish. Although reluctant to do so at first, the commission eventually recommended the protectionist tariffs, only to have President Eisenhower reject them in 1954. Fish stick processors welcomed the news, but vessel owners and labor unions balked.[18]

In fact, little could be done to ameliorate the brutal reality facing New England's fishing operations: to offset declining revenues following World War II, vessels had to fish harder and processors had to make and sell more fish products. These economic pressures further weakened severely stressed fish stocks so that by 1958, standing stocks of haddock, redfish, and flounders could no longer meet industry demands. This structural problem underlying New England's modern fisheries amplified all the others. As Boston College economists Edward J. Lynch, Richard M. Doherty, and George P. Draheim warned, "Much of cost of production difficulties that have plagued the New England groundfishery may be attributed to the adjustments that have been necessitated by the relative abundance or scarcity of the raw material, the fish."[19]

It was an odd twist to trophic cascades, a natural phenomenon in which energy transfer is altered by the changing abundance of key species within an ecosystem. Ecologists have shown that when one species experiences a dramatic change in abundance, resulting changes in energy flow permeate the entire food web and can reconfigure it to favor some species over others. Thus, those cascading changes can affect species and trophic levels far removed from the species experiencing the initial change.[20]

The experience of New England's fishing industry in the 1950s makes it clear that humans as well as marine species were considerably affected. Economic, political, social, and cultural changes affecting the region in the 1950s and 1960s add weight to theories about the outward rippling consequences of environmental change. In effect, the economic imperatives of industrial exploitation linked humans more tightly to marine resources across ever larger ecosystems through the need to transform more and more organic matter and energy into salable commodities.

This means that organic matter and money can both be seen as forms of energy and power. Linked to the marine food web through the transformation of marine organisms into commodities, New England's fishing industry suffered economic and ecological shock as local stocks drastically declined. Although the industry was suffering from international competition, labor unrest, bankruptcy, and rising production costs by the early 1960s, the effect of collapsing stocks cascaded through fishing communities as much as it cascaded through the marine ecosystem.

Acknowledging Decline and Laying Foundations for the Future

Against this somber backdrop, leaders from the heavy trawler fleet, including Thomas Rice, Patrick McHugh, Thomas Fulham, and others, actively engaged in developing major national fisheries legislation during the 1950s. To do so, they operated as sophisticated a lobbying program as any other modern heavy industry in the United States at that time. Unlike previous regulatory efforts, however, nothing quaint, traditional, heritage-laden, or romantic shaped what leaders wanted for their constituents. Like other professional lobbyists, they worked within Washington networks to obtain direct and indirect government subsidies. Hearing transcripts between 1954 and 1961 detail the complex, long-term strategy developed by leaders of New England's heavy trawler industry for seeking public assistance to save their industry. From the 1954 hearings over the Saltonstall-Kennedy Act, through major fisheries bills in 1956, 1958, and 1959, New England's vessel owners' association and labor unions recruited new allies among legislators and the public. They needed them. Almost all parties realized that harmful tariff decisions, changes in fish processing, and expanded foreign imports had put tremendous pressure on the industrial fleets. Equally apparent was how localized depletion of important fish stocks had exacerbated the chronic economic troubles that regularly sent industry representatives to Washington.

Industry leaders and legislators quickly articulated the major problems of the fisheries in a broad and comprehensive manner. Working with AFU leadership and representatives of the vessel owners, Massachusetts senators John F. Kennedy and Leveret Saltonstall sponsored a bill in 1954 calling for the use of revenues from imported fish tariffs to support programs enhancing domestic fisheries' competitiveness. Additionally, and with remarkable foresight, the bill called for a fisheries research program that placed resource stewardship and management on par with economic performance and competitiveness. Developed by the New England Fisheries Committee, an association of industry leaders and state fisheries directors recently formed to influence policy decisions, the proposal identified exploratory fisheries development, technological development, and education and market research as initiatives that should be supported with federal funds.[21]

However, the New England Fisheries Committee predicated their industry support initiatives on two fundamental proposals. Industry leaders called for biological and oceanographic research and improved statistical record-keeping in advance of new stock and gear development. In effect, this was tacit acceptance that commercial fishing had played a key role in depleting New England's commercial stocks. Thomas Rice, representing the New England Fisheries Committee, declared that "with or without regulation, New England haddock stocks are in a precarious position. Fishing has been too intense over the years and there is evidence of climactic change which may have a serious effect on this fishery." Redfish, too, needed scientific attention: "Research on this species is designed to determine how long the North Atlantic stocks can withstand [the] present rate of exploitation and to determine how, if ever, the nearby New England stocks can be restored." Likewise, newer fisheries needed research to prevent their demise. "Research on those valuable species [of flounders]," Rice continued, "is designed to protect against overfishing and the depredation of the trash fishery." Even scallops—a profitable fishery opened by New Bedford in the 1930s—needed attention: "The valuable New Bedford scallop fishery has been maintained by the fishermen moving from bank to bank as individual beds are fished out. How long can this go on? Research on this species is designed to determine how new beds are formed, how fast the shellfish grow, and how much fishing they can withstand." Finally, in perhaps the most forward-looking of all the proposals, Rice called for research on the effects of otter trawling on benthic habitat, noting, "In inshore waters there is a conflict of interest among the commercial ground fish, lobster, and sport fishery. The question of the effect that dragging has on the bottom, on lobster operations, on sport fish in nearby grounds and on other species of fish, is often raised." While an ulterior motive may have been excluding smaller draggers from competing with larger vessels by limiting inshore operations, this question proved so important to the New England Fisheries Committee that they called for designating an eighty-foot research vessel solely to carry out investigations.[22]

Initially created by state directors and vessel owners, the study program was quickly adopted by labor leaders. Patrick McHugh and the AFU agreed with its need: "To the fishery industry as a whole this problem of excess supply in certain instances and an undersupply in other calls for immediate serious study. Anything that can be done from a technical and biological standpoint should be done to help minimize as much as possible the various problems of the industry." In fact, the need for such a program was greater than ever before. Not only were cheaper foreign imports undercutting the market but "the depletion of much of our fishing grounds [has] caused this industry to receive several million dollars less for its product during the year 1953 than it did for any of the previous several years. . . . [New England fishing] is faced with more serious problems than at any time in its history."[23]

State fisheries directors echoed concerns about the economic consequences of depleted local stocks. For them, marine fisheries exemplified the need to extend the same principles of wise use and management to the marine environment that were beginning to restore game and freshwater fish stocks and the forests and rivers that supported them. According to the Massachusetts director of marine fisheries programs, Francis Sargent, "The New England fishing banks have been extensively harvested for some 300 years and the virgin abundance, as a direct result, has been reduced." Stanley Tupper, from Maine's Department of Marine Fisheries, laid out these lessons from history more extensively:

> Nature has provided man with a bountiful supply of resources to meet his every need. Wise men, even primitive men, knew that they must live in harmony with the nature in order to survive. He knew that, if he violated nature's laws, he would perish just as surely as other animals.
>
> However, history records that, as man has become more civilized, he has neglected to realize at times that food and shelter and clothing,

which are the basic needs of humans, still come out of the forests and fields, the waters, and the soil. In many instances, natural grounds of fish have been unthinkingly depleted or destroyed.

It is only recently that civilized man has become aware of the value of natural resources or has realized that an abundant life depends upon the preservation and intelligent use of these resources....

While the production by the American fisheries has increased, I would like to emphasize that this production has been maintained at high levels only by substituting new and, in many instances, less desirable species for old, by developing new market products such as packaged and frozen fillets, and by increasing mechanization of methods of capture. Every expansion of markets, every technological improvement in methods of processing fish, and every increase in the efficiency of the methods of capture has intensified the strain on fishery resources.

These economic and technological advances are destructive to the very fisheries they seek to develop unless accompanied by effective controls, management and research.[24]

While Tupper expressed concern about the economic future of Maine's fisheries, his support for the bill's initiatives rested on the greater need for responsible use and management: "Granted that civilized man has destroyed much of what was given to him, it is not too late.... Man's experience with land has taught him finally that without proper management of basic resources the harvests dwindle to a bare subsistence. The same can be said of the fisheries." To drive his point home, Tupper quoted the nineteenth-century British writer John Ruskin: "God has lent us the earth for our life. It is a great entail. It belongs as much to those who are to come as to us, and we have no right by anything we do or neglect, to involve them in any unnecessary penalties, or to deprive them of the benefits which was in our power to bequeath."[25]

Indeed, almost every supporter of the House bill acknowledged that depletion of New England haddock and redfish stocks undermined competitiveness. As veteran Essex County, Massachusetts, congressman William H. Bates put it, "Nearby fishing grounds have become depleted and as a result, the fisherman finds it necessary to travel four and five hundred miles before he can locate fish." Bates refused to attribute local depletions to decades of intensive fishing, however. Instead, he ascribed radical changes in stock health to causes "unknown."[26]

In fact, elected politicians on the New England Fisheries Committee such as Bates refused to look to the future as the industry representatives did. Nor did elected officials consider the present. Instead, Bates justified his support by invoking mythic nostalgia, which had been vital to fisheries debates before World War II. Bates argued that federal support for scientific studies, market research, and processing innovations would help support the oldest industry in the nation, one that trained seamen for naval service, produced food during national crises, and, with no apparent irony, had never asked for federal handouts. In Bates's testimonial, New England fishermen were not asking for subsidies, unlike many other industries at the time. Using imagery from the 1920s, Bates declared that fishermen were "merely seeking a guiding hand in the very specialized fields of techniques and science." Consequently, "the importance of the fishing industry to every man, woman, and child in this Nation is very apparent. In peace times and in war, it is invaluable."[27]

Nor was Bates alone in using nostalgic representations of, and paternalistic attitudes toward, New England fishermen. Washington congressman Thor Tollefson helped Bates out during the question-and-answer period. Somewhat disingenuously, Tollefson asked, "Is it not a fact that the fishing industry is perhaps one of the oldest industries in the United States?" "Perhaps oldest in the history of the world, along with hunting," was Bates's reply. Thomas Lane of Massachusetts carried the trope even further. Invoking imagery likely borrowed from Samuel Eliot Morison, Lane argued that "the seaports of Newburyport and Gloucester, and the self-reliant people that they bred, were once famous around the world, but now the ships and nets are idle, and the crews are trying to find part-time jobs on land." Lane even went one step further than Bates in ascribing a cause to the economic hardships that had befallen New England ports: "Men who look ahead say that we may, eventually, have to depend upon the oceans as the major source of our food supply. Men who live in the present[,] comparing that prediction with the run-down condition of the commercial fishing industry of the United States today, wonder how we can be so improvident." Again echoing arguments from the 1930s, Lane concluded that federal neglect of the fisheries caused the industry's degradation. Compared with federal support for agriculture, "Congress itself in my estimation has neglected the welfare of the fishing industry."[28]

Congressman Millet Hand of New Jersey; Francis Sargent, director of Massachusetts's Division of Marine Fisheries; and Stanley Tupper, commissioner of the Maine Department of Sea and Shore Fisheries, echoed this line.[29] Throughout the hearing and around the committee table, members of Congress joined the refrain that "the farmers of the sea" were entitled to as much support as farmers of the land. No one pointed out the fundamental difference between the two: that through land ownership, farmers were individually accountable for abusing their land, while fishermen, harvesting a public resource, were not. Such ideological challenges were rare in hearings during the 1950s, since the comparison provided political justification for supporting the bill.

The premise that the federal government neglected the fisheries—a statement ignoring almost six decades of federally funded research and vessel construction support—became central to the presentation of the bill. It was orchestrated in that way to provide Congress with a clear mandate for passage despite likely opposition from agricultural states. What was different this time from earlier arguments invoking New England's fishing past and its romantic imagery was who placed these things in the public record. During the 1933 Sail on Washington, fishing industry representatives argued before congressional committees that government had neglected them—with limited results. This time, industry representatives worked with congressmen and senators to craft the various bills that would become the Saltonstall-Kennedy Act, and they handed over the job of shaming the government to those within it. With congressmen presenting the social and moral arguments for supporting the industry, industry representatives could focus on the economic, regulatory, and policy aspects of the legislation that would give them what they wanted.

When the Saltonstall-Kennedy Act passed, it appeared that the fishing industry's strategy had worked to some degree: the industry received some of what it wanted. In the most important gain, public funds from tariff revenues were reallocated toward fisheries research programs. Awarded through a competitive grant process, Saltonstall-Kennedy subsidies could now support experimental fisheries investigations that benefited the industry. For the most part, the research sought to identify new fish stocks for commercial development and created new marketing campaigns boosting fish consumption. The Saltonstall-Kennedy Act did not include a wide-ranging ecological research program, however. Although such programs were fundamental to the future success of the industry, their results seemed too far removed from the docks in time and space for politicians to keep them in the bill. Still, the 1954 Saltonstall-Kennedy Act provided crucial economic relief for the industry, even if it ignored underlying ecological problems.[30]

Economic Expedients and Ignoring Decline: The 1956 and 1958 Regulatory Campaigns

In federal fisheries hearings after 1954, John Ruskin and regional history rarely appear in testimony. Moreover, hearings in 1956 and 1958 presented very little information about the biological obstacles New England's fisheries faced. Fish and Wildlife Service biologists, state marine fisheries directors, vessel owners, union leaders, and committee members ignored regional stock depletion when addressing the fallout from New England's fisheries collapse in the early 1950s. On rare occasions when the issue did come up, committee chairs stifled discussion. Between 1956 and 1961, deliberations over federal support for New England's heavy trawler fleet exhibited two striking trends. First, discussions from both the public and government focused on the economic needs of industry, intentionally rerouting deliberations away from the biological issues discussed in 1954. Second, representatives from the heavy trawler fishery began to solicit support from allies in other sectors of the groundfish fleet. Increasingly, seasoned lobbyists were joined by representatives from grassroots fishing organizations tied to small- and medium-vessel fleets, who were breaking long-established traditions of disengagement. In 1956 and 1958, after learning the ins and outs of Washington politicking from their large-scale colleagues, these organizations effectively performed important roles in fishing industry presentations to Congress. By the 1960s, such experiences helped redefine New England's groundfish industry.

Furthermore, deliberations in the late 1950s paid scant attention to the biological condition of New England fish stocks; instead, committee members and industry lobbyists focused intently on the economic challenges facing the industry. In 1956, Congress deliberated bills to offset industry production costs, subsidize vessel construction and refitting, and improve domestic fish processing. In addition to economic relief, those bills also considered a sweeping provision creating a national fisheries commission that would exercise powers held by the State Department and the Departments of Agriculture and Interior, which housed the Fish and Wildlife Service. Industry had urged federal support for fisheries research during the Saltonstall-Kennedy hearings, but in 1956, leaders of New England's heavy trawler fleet sought regulatory autonomy. Had Congress passed all the provisions presented by industry, a handful of lobbyists and industry leaders would have enjoyed unprecedented influence over federal policies and regulatory autonomy far greater than any other industry harvesting a public resource.

With so much at stake, it is not surprising that industry leaders mounted a concerted campaign to see their proposals successfully through committee. At the heart of this campaign stood the sympathetic legislators, committee members, and nongovernmental allies who supported the industry's positions. Even more than in 1954, these allies relied on the romantic imagery of New England's fisheries to underpin their arguments. The committee chairman, Senator Warren Magnuson of Washington, opened the proceedings with a sympathetic nod to an old refrain: "Generally, we have been talking in the Senate about farmers for a long time: there has been, in my opinion, a flagrant neglect on the part of the Government and all of us in the matter of farmers of the sea." Maine senator Frederick Payne also invoked timeless fisheries themes, this time ruefully acknowledging how threadbare they had become: "In the past few months I have so frequently made the statement that fishing is one of our oldest and proudest industries that I am hesitant to use the phrase again, but I believe it is worth repeating." He also reiterated New England exceptionalism in signaling his support for the bill: "For years now our commercial fishermen have been working with no more than the world-renowned Yankee ingenuity to help them. It is to their everlasting credit that our commercial fishing industry is in as good condition as it is today."[31]

This time, however, members of Congress outside New England voiced the longstanding arguments at play in New England's fisheries. Senator Thomas Kuchel of California stated, "Such legislation [as this] is required to revive the fishing industry from its present status of being almost an orphan, nationally, into high stature that it does deserve as a highly important part of the American economy." Robert Bartlett, delegate from the territory of Alaska, similarly alluded to the need for Americans to respect the fisheries: "The establishment of a commission and a sound and comprehensive national policy [as called for in the bill] would give the fishery industry an importance, a stature, and a dignity in a governmental sense that it does not now possess."[32]

It was no coincidence that committee members adopted the same old arguments. In addition to the New England Fisheries Committee, which still influenced fisheries legislation, Massachusetts fishing representatives forged alliances across the country to push their regulatory agenda. Thomas Rice, long affiliated with the Federated Fishing Vessels of New England and New York and the Massachusetts Fisheries Association, used his position in those organizations to make nationwide alliances aimed at influencing Congress on fisheries matters: "My association has worked, and will continue to work, very closely, with the tuna-boat operators of California in a joint effort to have these measures enacted by the Congress during this current session."[33]

The 1956 hearings witnessed a dramatic increase in the number and variety of organizations testifying in person or in writing. For the Saltonstall-Kennedy hearings two years prior, twelve civil servants and nine elected members of Congress joined thirteen industry lobbyists (none of whom came from Gloucester or New Bedford) to address the bill's provisions. In 1956, forty-six different lobbying entities (including six from Gloucester and one from New Bedford) provided input in person or in writing. Furthermore, discussions attracted lobbyists from beyond the usual suite of industry representatives. Fishermen's cooperatives, fishermen's wives associations, fishermen's unions, longshoremen, and seafood processing representatives from across the country spoke before the committee in March 1956. This influx of grassroots representatives reflected a growing politicization of New England's fishing industry beyond the established leadership of the heavy trawler fleet.

Along the way, the heavy trawler industry gained important new allies and provided their allies with tutorials on fisheries politicking.[34]

What newcomers witnessed was a deftly managed series of hearings. Magnuson used his role as chair to support his allies and marginalize views with which he disagreed. For example, industry representatives were permitted extensive, uninterrupted time to present their positions. Magnuson encouraged testimony by state fisheries directors who were sympathetic to industry's concerns. When discussions turned to the biological foundations of the national fisheries crises, however, Magnuson effectively shut down discussions of overfishing or restraints on fishing pressure. He openly sparred with Pennsylvania senator James Duff, who attempted to make conservation and reducing fishing pressure central to the discussion. In another instances, Magnuson jockeyed with George Difani, legislative representative for the California Wildlife Federation, who argued for the sound management of public resources.[35]

Magnuson kept the committee focused on the economic challenges fishing operations of all sizes and locations faced in the mid-1950s. In fleshing out these challenges, and providing allied committee members the political cover they needed, representatives from smaller, local fishing operations played critical roles. Gloucester's delegation featured prominently in the 1956 hearings. A lengthy sequence of statements—beginning with Massachusetts congressman William Bates, then followed by Gloucester's mayor, a city councilman, and representatives from the newly formed Gloucester Fisheries Commission—portrayed the hard times wracking the port. Each delegate laid out how insurance costs, lack of credit, low prices paid to vessel owners, labor stoppages, and international competition undermined the viability of Gloucester's fishing industry. Indeed, part of the delegation's strategy rested on the power of personal testimonials. Solomon Sandler, secretary of the Gloucester Fisheries Commission, stated as much: "In my opinion, the only ones who can express the real feeling, the intimacy of the Gloucester waterfront, are those people who are financially and physically connected to it."[36]

Gloucester's delegation were no puppets, however. Sandler made it very clear that his group would not fall in line behind lobbyists from the heavy trawler industry: "Any witness who comes down here and speaks as a representative of a group represents a particular group only, and in no event does he speak for the entire industry." This was an important point. While Boston's heavy trawler representatives avoided mentioning depleted local stocks and the resulting increase in operational costs, Gloucester's representatives were determined to present the biological problems central to their arguments despite Magnuson's objections. Gloucester councilman John J. Burke stated, "We have to steam from 600 to 1,000 miles before we can catch our fish, which makes our costs much higher [than Canadian producers]." Joseph M. Cody, a member of the Gloucester Fisheries Commission, agreed: "Where [Canadians] are close to the banks they make a round trip in 3 days, 4 at the most, and they can do that. Compared to, up here, we make a trip from 8 to 10 days.... You couldn't even come close" to matching their price. Gorton-Pew's representatives also made depleted local stocks of cod, haddock, and redfish a focal point. "When [local fishing firms] stopped landing fish which could be used [for producing salted fish]," the spokesperson noted, "we imported them and processed what we could in Gloucester, but a number of years ago, we saw a tendency for the North Atlantic species to be caught and landed nearer the banks in Canada." As a result, Gorton-Pew moved their operations to Canada: "We are up there because we can get a supply of fish."[37]

Ultimately, the proposals presented in 1956 called for carving out of preexisting agencies the authority to manage marine resources and relegating those powers to a national fisheries commission filled with industry leaders. Not surprisingly, the proposals failed to pass, mostly due to the legal and jurisdictional problems a national fisheries commission would create. What emerged from the 1956 legislative session, however, was a call to establish a national policy highlighting the unique and influential role fisheries played in the national economy.[38] More important for the New England fisheries industry were benefits that fell outside regulation. The 1956 hearings had expanded New England's fisheries constituency and mobilized a wide array of fishing organizations representing a broader cross-section of the fleet.

In subsequent 1958 hearings, those constituents and interests were better organized, more united, and ready to do battle. By then, as economics Lynch, Doherty, and Draheim would reveal three years later, New England's groundfish fishery was already suffering from crippling economic losses due to stock declines. That year, however, industry lobbyists reached a dénouement in the tariff campaign they had begun in the early 1950s. Leaders of heavy trawler fleets had previously appealed to the U.S. Tariff Commission for protections against foreign fish products entering the United States as part of the Marshall Plan's aid to Western Europe. After initially splitting its votes on whether to recommend protections, the commission unanimously supported the groundfish industry in 1956. Yet President Eisenhower, locked in the heat of the Cold War and unwilling to alienate European allies, refused to raise tariffs on European fish. Over the next two years, prices and profits from the ground fishery fell all over New England. Boston's heavy trawlers offset falling prices by catching more fish on distant grounds. The sailing range of small and medium vessels was limited, however; they were stuck on New England's barren banks.

By 1958, New England's groundfish industry was ready to bring its case to Congress again. It launched a campaign to restore its industry against what appeared to be the federal government's desire to kill it. With the support of "90 percent of all the concerns presently engaged in the commercial fisheries industry of New England," Thomas Rice, now representing the New England Committee for Aid to the Groundfish Industry, presented to Congress a bill to subsidize insurance costs as well as vessel construction, secure government loans for shoreside modernization, and establish minimum prices paid to fishermen and owners.[39]

From the start, Rice made clear that the 1958 bill culminated a long, well-organized, carefully choreographed, politically engaged, defensive campaign. This legislation, he commented, "represents the climax in a long and bitter struggle, waged by the New England fishing industry, to regain for itself the right to live, prosper, and grow as an integral part of the industrial life of these United States." After falling victim to what they viewed as governmental abuses, deceptions, and false promises, industry leaders "found ourselves again on the merry-go-round to nowhere. It has been our fate in our dealings with the Federal Government to experience about every human reaction known to man."[40]

Despite government subsidies and aid for the industry over the previous four decades, members of the New England Committee for Aid to the Groundfish Industry stated, "We have been ridiculed, rebuffed, treated with abject indifference, and totally ignored. We have been received warmly and lulled with false promises of aid and assistance. Worst of all, we have been frustrated and victimized by the delaying and stalling tactics of officials in high places." Given his feelings about the matter, it is no surprise that Rice bullied Congress, demanding increased federal support to redress previous injustices: "The four points [in this bill] represented the absolute minimum in the way of aid and assistance needed to stabilize the industry. The acceptance of any single point or series of points, any action short of full and simultaneous enactment of all four points would not be sufficient to halt the decline or provide a sound basis for rehabilitation."[41]

To Rice, the industry was guiltless in terms of its current predicaments. In his words, "Our present deplorable condition is not of our making. The New England fishing industry was, in its day, a sound and healthy segment of our national economy." Ignoring three previous decades of federal investments into product development, market establishment, more efficient gear, and indirect subsidies, including significant federal funds spent on the research and development of unheeded recommendations to prevent local haddock and redfish stocks from crashing, Rice portrayed a fishery victimized by governmental indifference: "[The U.S. government] constituted a force so overwhelming, so far reaching in its activities that the domestic industry was rendered impotent and powerless to compete against it." According to Rice, postwar economic policies had victimized New England's fisheries to an excessive degree. Much of the foreign fish flooding American markets not only enjoyed generous tariff rates but was produced by technology, equipment, and facilities built overseas by the United States during the war, then transferred to foreign ownership to help rebuild shattered economies. Rice reiterated what the industry had argued for years, that U.S. foreign policy put national interests ahead of the interests of New England's groundfish industry. This created a "series of impossible conditions imposed upon the industry by those in Government who would willingly sacrifice whole segments of our national industrial economy in order to perpetuate a theory that free trade between nations is the sole solution to our international problems." Echoing New Deal refrains, Rice concluded, "our desperate pleas for justice and equity have been completely ignored."[42]

Then, Rice laid out a scripted, well-orchestrated battle plan to secure what he and his supporters felt was their due: "The testimony presented by witnesses to follow will disclose in detail the obstacles and the conditions each segment of the industry encounters in the day-to-day struggle for survival. The testimony will illustrate clearly why each section of the bill now before you is essential and vitally needed if we are going to restore the industry to a normal position in the Nation's economy." Gloucester mayor Beatrice Corliss also made clear that the industry was organizing politically. She noted, "The continuation of the fight [for federal direct and indirect subsidies] led to the Gloucester Fisheries Commission joining with Boston, Portland, New Bedford, in forming the New England Committee for Aid to the Groundfish Industry in a fight for a permanent solution" to the industry's troubles.[43]

Gloucester's emergent lobbyists played a large role in Rice's strategy. Of the thirty-three lobbyists providing comments for the 1958 hearings, twelve hailed from Gloucester. In addition to the Gloucester Fisheries Commission, who, in person, attended the hearings in Washington, the public record included statements submitted by the Gloucester Vessel Owners Association, the Gloucester Chamber of Commerce, the executive vice president of the New England Council, and the Gloucester Fisheries Association, as well as several private citizens. Rice also reached out to allies from around the country, whose representatives added their voices in support of the suite of fisheries bills under consideration. In all, more than fifty lobbyists submitted comments on the five fisheries bills being considered in 1958, with some organizations commenting on more than one bill.[44]

To persuade the committees of the value of their position, Rice and select allies presented data-driven arguments to augment their credibility. Elected officials complemented this approach by presenting familiar cultural arguments to the committee that were now stock-in-trade at federal level New England fisheries hearings. Maine senator Frederick Payne made the historical argument, as he had done in earlier hearings: "Fishing is one of the oldest and proudest industries of our Nation. In fact, this country was really pioneered by fishermen who were followed by the colonists. From that early day to this fish has been, and properly so, a cornerstone of the American diet, and the industry has been an important segment of our national economy." Leslie Dyer, president of the Maine

Lobsterman's Association, made the farmer-fishermen analogy: "We are in reality, seagoing farmers harvesting the products of the sea, taking our own risks." Finally, Gloucester Fisheries Commission member William Cafasso presented the national security argument: "Our fleet, gentlemen, has been among the very first to answer every national emergency. It is a matter of record that the fishermen of this country have responded to every war with ships as well as a quick supply of food." Even John F. Kennedy, then senator from Massachusetts, submitted written comments presenting the traditional heritage argument in favor of federal fisheries subsidies. "It is apparent that fishing is not only the oldest industry in New England," he stated, "but one of the most important."[45]

Most striking in the 1958 legislative discussions, however, was the almost universal denial of the biological distress of New England's groundfish. Whereas in 1954, both labor and management united to push for long-term conservation, research, and management measures, few in 1958 mentioned the biological health of the stock. Victor Turpin, Patrick McHugh's successor as secretary of the AFU, raised the issue briefly: "The fish are not coming, either, as they should."[46] Others only mentioned the topic to dismiss it as irrelevant. Donald McKernan, director of the Bureau of Commercial Fisheries, contradicting almost a decade of industry testimony, argued,

> There is another matter, of course, that some people have brought up from time to time, and that is that actually we have a rather limited fishery, and that we have overextended ourselves, maybe. Of course, this could be an argument in this case, but that is simply not the case. The general level of productivity of the fishing grounds in the northeast, off our northeast coast and the Grand Banks, the Georgia Banks, is relatively high at the present time. Conservation measures which have been taken as a result of the North Atlantic Fisheries Treaty . . . have produced remarkable results in recent years. . . . And these have been effective and they have actually increased the stocks of these fish on the grounds, rather than decreased them. So that there isn't any biological reason for [declining catches].[47]

These deliberations witnessed the arrival of Gloucester as a prominent new force in New England's fisheries leadership. In featuring detailed testimony from Gloucester's delegation, the bill's final report illustrated how much influence had accrued to Gloucester's leadership during the 1950s legislative hearings. With the large trawling firms moving overseas or folding, the leadership of Thomas Rice and the AFU was giving way to local fishing associations organized and operating in Gloucester.

More than Rice and the AFU, these local groups recognized the power that New England's traditional fishing iconography wielded in public hearings. In their desire to present the heavy trawler fleet as modern and industrially significant, Rice and the AFU underplayed the powerful imagery from the 1920s and 1930s. Indeed, rarely in the 1950s did these leaders invoke the romantic past, although they happily benefited from others doing so. Gloucester's leaders had no such mandate of modernity. With a port filled with vessels of all sizes and ranges, the whole town suffered from depleted local stocks and intense foreign competition. Gloucester delegates had no qualms about using their historical and cultural heritage to win policy concessions. That their fishing operations were as modern as the heavy trawling firms mattered little. If politicians chose to see them as quaint extensions of New England's bygone schooner fisheries, so be it.

In the end Gloucester's tactics gained some relief for New England's fisheries. The Fisheries Assistance Act of 1958 appropriated an additional $10 million to preexisting funding sources that provided assistance to fishing industries nationwide. The new funding would offset construction costs and help reduce operating and overhead expenses faced by American fishing firms.[48]

Romanticization and the Consequences of Stock Decline: The 1959 Regulatory Campaign

Within a year of Rice's 1958 efforts, he was again on Capitol Hill requesting more government subsidies for New England's fishing industry. This time he focused on the fleet's age and technological obsolescence. Arguing that New England needed new, more efficient vessels to compete against foreign imports, Rice orchestrated yet another campaign to secure direct federal subsidies for vessel construction. At the center of Rice's complaints stood a 1796 law requiring all U.S. fishing vessels to be built in the United States. This law prevented New England interests from building new vessels in low-cost overseas shipyards. In 1959, the House Committee on Merchant Marine and Fisheries once again heard testimony from New England fishing interests pleading for governmental aid, this time in the form of differential subsidies that would even out the costs of vessel construction in domestic and foreign shipyards.[49]

Again, New England's congressional delegation reminded the committee of the historical and cultural value of New England fishermen. This time, veteran Massachusetts congressman and House majority leader John W. McCormack laid out the lesson: "The commercial fishing industry of the United States, as you know, originated in the Commonwealth of Massachusetts, when the colonists first settled Plymouth, Boston, and Salem." McCormack, somewhat ambitiously, credited New England's fishing industry with settling the rest of the continent, allowing him to argue that "the fishing industry played an important role in the growth of America. Young New Englanders from the commercial fisheries of the area migrated to Lower California and established the great tuna fisheries of the Pacific. . . . With such a history, it certainly cannot be said that the citizens engaged in the fisheries of New England were remiss or neglectful citizens in managing their own industrial affairs." Instead, McCormack laid contemporary problems squarely on the shoulders of government: "The blame for the industry's present depressed condition lies solely with an outdated, antiquated tariff structure that has not been adjusted to coincide with our modern conditions." Consequently, it was incumbent for Congress to redress this problem, for "we must keep the ground fish industry of New England alive."[50]

Essex County, Massachusetts, representative William H. Bates followed McCormack's lead. A veteran of earlier fisheries legislation campaigns, Bates presented the human costs of indifferent governmental policies:

> Now, if it is national policy—and it is—that the shipbuilding industry of the United States should be subsidized [by mandating domestic fishing vessel construction], it seems to me that the Nation as a whole should bear the brunt of that cost, and not the poor, desolate, impoverished vessel owners. . . . It is unfair and unjust to expect the poor fishing vessel owners to pay that differential [in cost between foreign yards and domestic yards]. . . . And if we continue the way we are today, we will have

the complete destruction of the entire fishing industry that for 300 years has been a great industry.

In resurrecting images of fishermen's independence and autonomy, Frank M. Coffin from Maine completed the well-established picture of the New England fishing industry's plight: "No one likes subsidies and probably no one likes them less than the hardy individualists in our New England fishing industry. But faced very literally with the prospect of extinction, they have no alternative."[51] Again, New England delegation's painted a familiar caricature of New England fishermen.

Rather than a modern industry seeking its just rewards from an industrial economy and the national government, older visions of New England fishermen provided the most political cover for Congress to aid the region's beleaguered, inefficient, and embattled industry. The New England delegation in the House testifying on the floor wanted to make clear that governmental tariff policies had created the depression in the fisheries. Additionally, federal protections for domestic shipbuilding were preventing fishing firms from solving one of their current problems, vessel construction costs. Once again, the federal government must set things right for the venerable and culturally valuable New England fisheries.

When it was time for Gloucester's delegation to make its presentation, fisheries advocate Solomon Sandler, this time as a private citizen, recounted to the committee how previous hearings had carried Gloucester from hope to heartbreak: "When I spoke here last year I had a prepared statement which took us 2 years to prepare. . . . When the group from Gloucester and the group from Boston left this committee last year [however], it wasn't but a matter of 2 or 3 weeks before we all felt that it was an absolute waste of time coming down here." As a result, Sandler felt the committee owed them: "We are asking for some real consideration this time."[52]

Thomas Rice agreed and took the committee to task for a decade of frustrations. In a lengthy statement, Rice recounted how he and his allies had sought protections as early as 1947. Through the entire period, they felt the federal government had snubbed their demands. Eisenhower's denial of tariff protections, in particular, stuck in Rice's craw. It energized him to form grassroots lobbying organizations and mount the sustained campaign of recent years. Congress, in Rice's view, had also dealt unjustly with his industry by not bringing bills to the floor for a vote that he had drafted and submitted to various committees. In short, Rice all but demanded that the committee make recompense in 1959 for denying New England's fishing industry special consideration. An earlier version of this bill, Rice observed, "was worded in such a manner that only New Englanders could be the recipients of its intended benefits. It was so worded at that time because no other segment of the commercial fishing industries of United States had ever experienced the debasement, the humiliation, the indifference, the frustration, the lack of cooperation and understanding in dealing with the US Government as has the New England fishing industry. . . . We suffered these inequities and injustices in pursuance of an established principle of democratic procedure."[53]

More than anyone, Gloucester Fisheries Commission and City Council member William P. Cafasso articulated Gloucester's indignation: "It is almost ludicrous, in fact, when we think that here we are coming to Washington, year after year, pleading for some kind of assistance when actually we are not the type of people who are looking for a handout but we do want to be placed on competitive basis, at least, with the rest of the world." That humiliation, combined with the large number of trawlers placed into wartime service, sold off to foreign countries, then exporting cheap fish products back to the United States, convinced Cafasso that the federal government owed Gloucester and New England assistance.[54]

Not everyone bought the argument, however. John Dingell of Michigan challenged Rice by asking, "You are catching in some areas immature fish, are you not, because you are fishing in certain areas instead of going far enough abroad to explore new fishes like, let's say, the Japanese and Russian and English." George Miller of California also raised doubts about the New England contingent's assertions of indifference and victimization. Commenting on Cafasso's claim of government discrimination after World War II, Miller stated,

> I sometimes get a little perturbed when you introduce arguments that try to leave the thought that fishermen and fish[ing] vessels owners were discriminated against as a result of the war. As a matter of fact,

they got better prices for their vessels than they could have sold them for on the open market. Now then, if some man making a good profit on his vessel, because he sold it to the Government in time of war did not want to go back into the fishing industry, I do not think it is becoming to come here and complain that the Government was responsible in part for taking [away] his vessels.[55]

Others, even in New England, doubted whether new vessel subsidies would solve the problem. Hy Trilling, president of the heavy trawling firm Boston Bonnie Fisheries, testified that "it is common knowledge that our New England fishing industry is relatively dead.... In fact, we are already dead but refuse to bury the body. I say further, better an industry be fully dead than partially alive." Furthermore, because Boston auction houses set national prices, anemic volumes of landed fish artificially inflated prices across the country: "This little bit of fish produced in New England is costing the American public many millions of dollars per year." Donald McKernan, this time representing the U.S. Fish and Wildlife Service, drew from statistical analyses to question whether foreign competition was truly to blame at all. He noted, "The real problem that has been out before this committee is that it costs so much to produce fish in New England; it costs so much to produce fish here that we cannot compete with the costs of production in these foreign countries."[56]

Unlike in 1954, when stock health was openly discussed, or in 1956 and 1958, when repeated attempts to discuss stock health were squelched by the committee chair, few raised the issue of the region's ecological productivity at all in 1959. The bill simply questioned the proper role of subsidies in American industrial productivity within a larger global market. Yet questions of stock health still crept into view, desired or not. Rice pointed out that in addition to being a victim of government, "for the past three years we have been a victim of nature itself in that the amount of fish available on our local banks has decreased considerably." To him, this was part of a natural cycle of fisheries abundance and scarcity that, according to Fish and Wildlife Service projections, "would rectify itself in 1960."[57]

After five days of testimony, delegations from Boston and Gloucester had painted a clear picture for committee members to consider. Federal tariff policies and technology transfer programs were undercutting the

American fish market. The only way to compete was to catch more fish more efficiently. With thin or nonexistent profit margins, a victimized New England fishing industry could not afford to build new vessels in the United States to do so. Consequently, it was entitled to government subsidies for vessel construction to offset the difference between domestic and overseas production.

This time industry won. In response to the haranguing, the romantic imagery, and the guilt and shame lavished on the committee, Congress enacted a three-year vessel subsidy program, with the federal government covering one-third of the cost of constructing new fishing vessels in American yards. The 1959 hearings marked an important turning point in how New England fishing interests would campaign for federal support and how New England fishing would be defined. In orchestrated committee presentations, Corliss, Cafasso, and Rice trotted out the same images of the neglected, ignored, and marginalized fishermen that Gloucester had previously used in the nineteenth and early twentieth centuries. The tactical shift reflected the realization that a decade of attempts to secure federal support based on the modernity and efficiency of New England's heavy trawlers had mostly failed. Instead, the few gains won in the 1950s came when legislators saw the fishery as traditional, timeless, and victimized.

Such images of victimization were not merely representations: Eisenhower's tariff policies had brought in imports that undercut American market prices. While good for American consumers and U.S. foreign policy goals during the Cold War, those policies challenged the New England fishing industry, small- and large-scale operations alike. Together, the market realities and political representations framing New England's fisheries in the 1950s pulled Gloucester and its fleet to the fore of fisheries lobbying. As a result, Gloucester's leaders took on greater roles in campaigns to wrest subsidies from the government in the face of challenges both real and imagined.

Throughout this period, Rice orchestrated a sophisticated campaign presenting the historical, cultural, social, and economic arguments for greater federal aid to the region's fisheries. Since the early 1950s, he had expanded his base beyond Boston's heavy trawling interests by drawing on new local fisheries associations representing other sectors of the groundfish fishery. Tied to redfish, less unionized, and with a diverse fleet structure, Gloucester provided new, powerful, and sincere voices that resonated in the hearing room. Marshaling his resources, Rice transformed how the industry related to government by training those new voices in a rough apprenticeship of fisheries politicking.

By the late 1950s, the AFU's new secretary, Victor J. Turpin, struggled to make sense, even to explain clearly, how fishing had come to mean much more than catching fish. He testified that "the fishing industry was a noble industry at one time.... This industry was fine when the fishermen themselves were operating it for a livelihood. When people began investing money into for benefit [sic], commercialize, and so forth, it is a vast difference between me making a living in the fishing industry and having money invested, loans, as it were, as benefits. Investments they look for, and they look for interest on their money. This industry has come a long way from manual labor."[58] Turpin was struggling with the simple reality that with the commercialization of the fisheries came their politicization, and this fact emerged more clearly by 1959 than ever before. Although smaller vessels had long deferred to the political leadership of New England's heavy trawling industry, by 1958 the ecological foundation of that political authority had waned. Small- and medium-sized fleets embodied a cultural power that Rice had not fully utilized. As different sectors teamed up to fight for greater government subsidies, the increased breadth, sophistication, and orchestration of committee hearings reflected a new fundamental truth about New England groundfishing: politics and popular perception were as important to fishing as catching fish.

Rice and the AFU failed to win federal concessions not merely due to political adversity and federal Cold War policy but also due to a reallocation of political power. The impact of local stock declines cascaded through local fishing communities in the form of labor unrest and dwindling profitability, and continued to intensify political tensions. Throughout the 1950s, federal hearings showed that Boston's haddock fleet was losing political influence. Beyond receiving vessel construction loans, which aided larger firms, Boston fisheries' protracted campaign for tariff relief yielded few benefits to other segments of the industry. Nor had it stemmed the tide of imported fish products, which many in the industry believed was undercutting American markets.

The waning political influence of the heavy trawler fleet also stemmed from ecological changes. Depleted local stocks undercut industry profitability, which undercut its political influence. Degraded stocks, high overhead costs, and the inability to generate meaningful profits all stemmed from the high costs of finding full fares. Under these circumstances, it was difficult to argue that this fleet was a modern, efficient, industrial success, as Rice had tried to do. If declining stocks caused labor unrest and industrial outmigration earlier in the decade, they also undermined the political leadership of Boston's haddock fleet. Ecology, economy, and politics were all linked together in a downward spiral.

The 1959 hearings also established that the broader grassroots movement, which proudly claimed to be traditional and artisanal, advocated more effectively. Gloucester's lobbying delegation rose to prominence in this campaign by arguing that its city's fishing industry had long been victimized by callous and indifferent federal policies. In doing so, Gloucester's representatives redefined what a New England fishery looked like. Resurrected images from earlier in the century presented New England fishing as small in scale, docile in industrial relations, humble in demeanor, accepting of providence, and—importantly—victimized by forces beyond its control. In the future, popular perceptions of New England fisheries as local, artisanal, and small in scale would play a more prominent role in federal lobbying efforts. This shift may seem insignificant, but it would shape the future of New England fishing for the rest of the twentieth century.

CONCLUSION

In one sense, Gloucester's sense of victimization rested on solid truth, though not perhaps the truths Gloucester's leaders believed in. While their legislative campaigns were waged against foreign imports, and later foreign fleets, the real issue to which almost every New England fishery fell victim eventually was decline in commercial resources. Throughout the northwest Atlantic, haddock and redfish stocks, the bedrock of New England's heavy trawler fisheries, could not withstand the unrelenting fishing pressure to which the New England fleet subjected them—even before foreign fleets began targeting haddock in the early 1960s.

Fisheries data reveals the extent to which a declining natural resource base restructured New England—and in particular Massachusetts's—fisheries in the 1950s and early 1960s before foreign fleets arrived in earnest in 1963.[1] Reaping windfall profits in 1946 targeting mostly haddock and redfish, by the late 1960s otter trawl fisheries could no longer efficiently catch fish. Congressional testimonies and federal records support this grim conclusion. Landings per day absent (LPDA), defined as the landings generated for the length of time a vessel was away from port, for the entire Massachusetts fleet working throughout the northwest Atlantic fell precipitously for redfish and haddock over two decades (fig. 2).

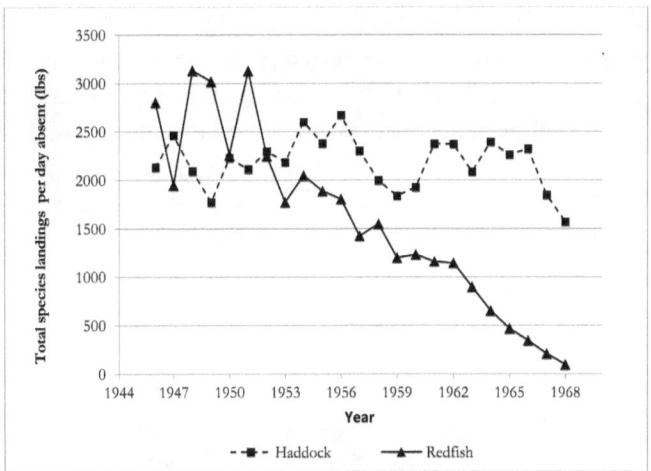

FIGURE 2. Pounds of haddock and redfish LPDA, Massachusetts otter trawlers, 1946–1967. Data from: Annual reports in the U.S. Department of the Interior, Fish and Wildlife Service, "Landings at Certain Massachusetts Ports," in *Fishery Statistics of the United States*.

Declining efficiencies, in turn, changed the structure of the New England fishing fleet. The size of the otter trawl fleet and each vessel's annual days absent from port similarly declined in response to the economic pressures the industry faced (fig. 3).

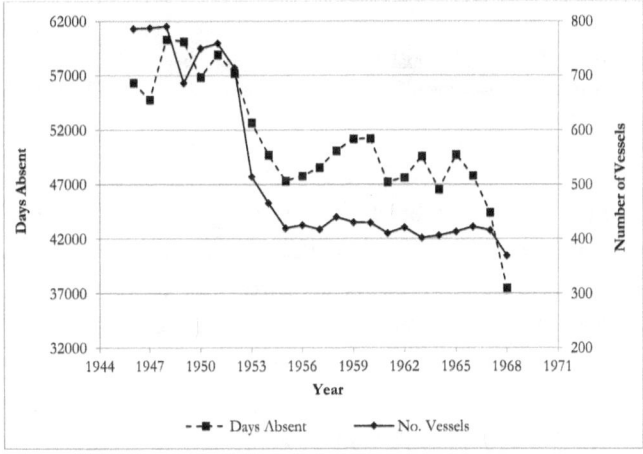

FIGURE 3. Total fleet size and annual days absent per vessel, Massachusetts otter trawlers, 1946–1967. Data from: Annual reports in the U.S. Department of the Interior, Fish and Wildlife Service, "Landings at Certain Massachusetts Ports," in *Fishery Statistics of the United States*.

The consequences of sustained otter trawling, however, came home to roost most dramatically in New England's own back yard. Data from fishing activities in ICNAF Area XXII, comprising New England and Georges Bank, reveals a fleet struggling to maintain catches in the face of declining productivity. Fewer vessels worked New England waters over the period, but those that did so worked harder, as defined by days absent per vessel per year. As the fleet dwindled, remaining vessels fished harder to find elusive prey, adding further strain on men and boats (fig. 4).

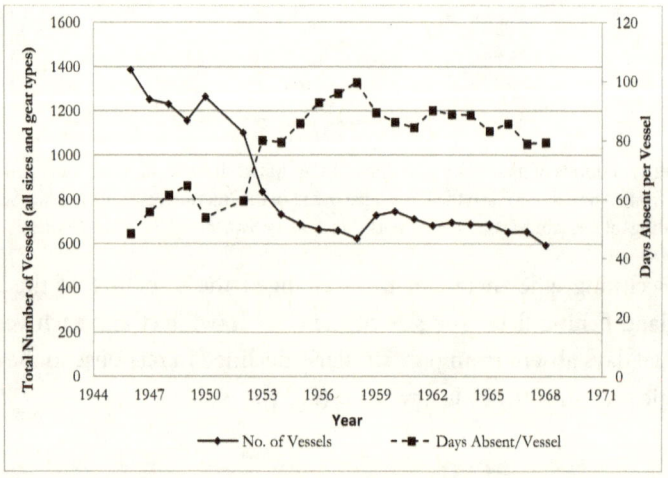

FIGURE 4. Vessels fishing and annual days absent per vessel, New England area, 1946–1968. Data from: Annual reports in the U.S. Department of the Interior, Fish and Wildlife Service, "Landings at Certain Massachusetts Ports," in *Fishery Statistics of the United States*.

By the time Boston lobbyists were arguing against President Eisenhower's tariff policies and Gloucester's leadership was claiming victimization at the hands of foreign fleets, the real damage to New England's small-vessel fishery had already been done. Between 1952 and 1953, the number of small otter trawlers operating in Massachusetts plummeted from 476 to 282—almost a 50 percent drop from the fleet's 1952 peak of 524 vessels. Other fleets also shrank due to local scarcity: the medium otter trawler fleet fell from 238 in 1946 to a low of 143 in 1955, but grew again to 163 vessels by 1961. Still, that fleet represented only 60 percent of its peak size. The heavy trawler fleet fared the worst. Even before the fleet-wide landings collapse of the mid-1960s, Massachusetts's heavy trawler fleet fell by more than half, from a 1948 high of eighty-five vessels to just forty-one in 1961.

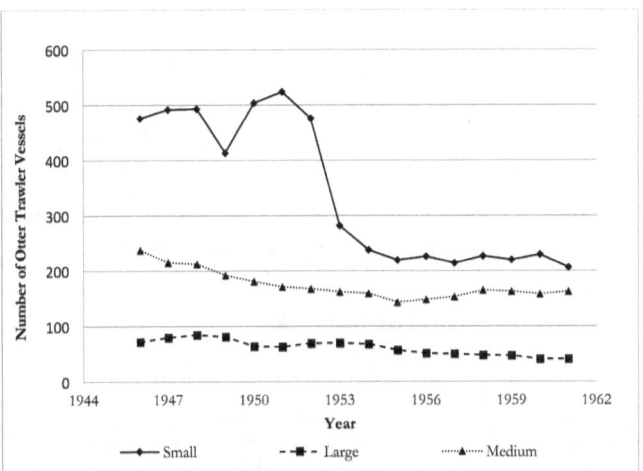

FIGURE 5. Size of the Massachusetts otter trawler fleet, 1946–1961. Data from: Annual reports in the U.S. Department of the Interior, Fish and Wildlife Service, "Landings at Certain Massachusetts Ports," in *Fishery Statistics of the United States*.

By the mid-1960s, declining stock abundance throughout the northwest Atlantic had fundamentally restructured New England's fishing fleet. Boston's heavy haddock trawlers—many just twenty years old—could no longer catch enough fish to pay operational costs. Turning enough profit to attract new investors, hire younger fishermen, or afford decent insurance seemed to be a pipe dream. The medium and small otter trawling fleets went the same way. One of the most efficient fishing methods ever seen, otter trawling took only forty years to undermine one of the most abundant fisheries the world has ever known.

Local Ecological Consequences

As New England's fleet underwent fundamental restructuring, New England's waters were experiencing its ecological equivalent. Key indicators of stock and ecosystem health suggested that local waters were changing under the pressure of otter trawling. As Michael Graham and E. S. Russell had shown in the 1920s, catch efficiencies not only indicate the economic viability of an operation; they can also roughly gauge the health of the stock that supports it. While a coarse metric, LPDA for local redfish and haddock stocks during the 1950s and 1960s suggest that fishing operations outstripped the ecosystem's ability to keep fishermen employed.

The haddock fleet had shrunk dramatically by the mid-1950s, and remaining vessels fished a lot harder. Initially, this strategy worked, and landings efficiencies increased with days fishing until 1953 (fig. 6). As each vessel spent more time fishing, however, landing rates declined until the mid-1960s, when New England fishermen and the foreign fleets redoubled their effort on haddock.

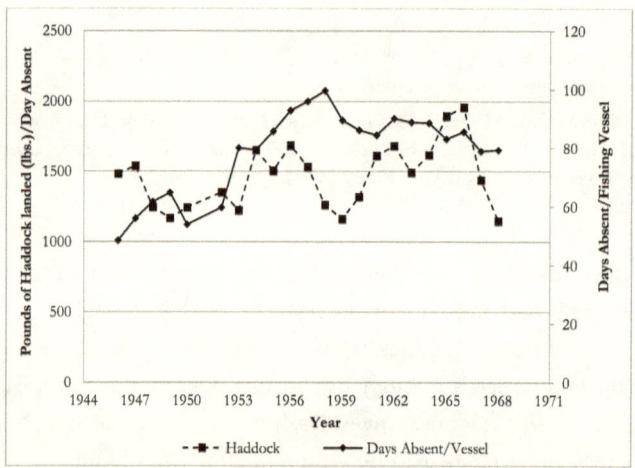

FIGURE 6. LPDA and days absent per vessel, all gears working in New England waters (ICNAF Area XXII), 1946–1968. Data from: Annual reports in the U.S. Department of the Interior, Fish and Wildlife Service, "Landings at Certain Massachusetts Ports," in *Fishery Statistics of the United States.*

Nevertheless, catch efficiencies fluctuated significantly (fig. 7). The New England fleet worked harder and targeted juveniles (scrod) to make up for decreasing landings of larger haddock. This happened at least twice in the postwar period. In the first instance, haddock LPDA in New England waters dropped from a 1947 peak to a trough that lasted through 1953. During that time, the percentage of scrod haddock in the catch increased markedly. Concerned with the health of juvenile haddock stocks, the U.S. Fish and Wildlife Service commissioned a study, "Destruction of Undersized Haddock on Georges Bank, 1947–1951," to assess the biological consequences of the fleet's fishing efforts.[2] But the proportion of scrod haddock declined between 1954 and 1957, and LPDA increased with greater catches of adult haddock. When large haddock landing rates fell again between 1958 and 1960, fishermen again turned to juvenile haddock to make up their fares.

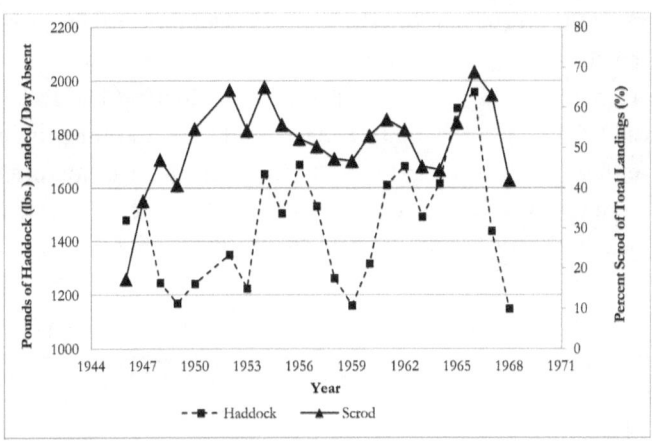

FIGURE 7. Haddock LPDA and percent scrod of total landings, New England waters (ICNAF Area XXII), 1946–1968. Data from: Annual reports in the U.S. Department of the Interior, Fish and Wildlife Service, "Landings at Certain Massachusetts Ports," in *Fishery Statistics of the United States*.

Changes in the ratio of scrod to large haddock can be explained by changes in the quantity of juvenile fish grown large enough to be caught in a trawl net. Clearly, this recruitment to the fishery affects effort, but the industry's focus on scrod haddock appears not only in contemporary landings data but also in stock structure studies examining historical stock size, composition, and fishing mortality rates.

According to a 1982 study, certain peaks in the percent of scrod haddock in total landings correspond to peaks in fishing mortality for age-1 and age-2 fish, back-calculated using current length-at-age conversion factors applied to historical data. Fishing mortality peaks and peaks in the proportion of scrod haddock to total catch occurred in 1948, 1950, 1952–54, and 1960. Correspondence is unclear for other data peaks because fluctuating recruitment and growth rates also influenced the abundance of scrod haddock. Nevertheless, fishing mortality among scrod haddock was elevated for six out of the thirteen years from 1948 to 1960, suggesting that the industry targeted juvenile fish during those six years. They were content to fill holds with whatever sized fish they could find.[3]

After 1963, greater landings of juvenile haddock briefly increased catch efficiencies again. Nonetheless, two years later, as foreign fleets came in to finish off what remained of New England haddock, New England fishermen drove up total haddock LPDA and the proportion of juveniles taken in their hauls. Fishing effort on immature haddock became so widespread that government data collectors created a new size category, "Snapper Haddock," to record how many pounds of tiny haddock the fleet landed.[4] Haddock were hounded by domestic and foreign fleets until 1968. Then the backbone of New England's twentieth-century industrialized fishery simply snapped.

For redfish, the consequences of sustained, intense fishing in New England waters proved even more dramatic. From a 1946 high of just over 1,200 pounds (0.5 mt) LPDA, local stocks fell off to roughly 240 pounds (0.1 mt) in 1953 and declined steadily to a low of 29.2 pounds (0.01 mt) LPDA in 1967. First commercialized in the mid-1930s, by the mid-1950s daily catch rates had plummeted in the fishery, and by the mid-1960s they had collapsed even further (fig. 8). Haddock landings suggest a different story on the surface. The resource appeared to yield a higher LPDA until the mid-1960s. Yet, as we now know, that trend masked the underlying realities fishermen faced after World War II. The decline of Georges Bank haddock was significant enough to launch federal scientific studies of long-term trends in the haddock fishery.[5]

CONCLUSION 175

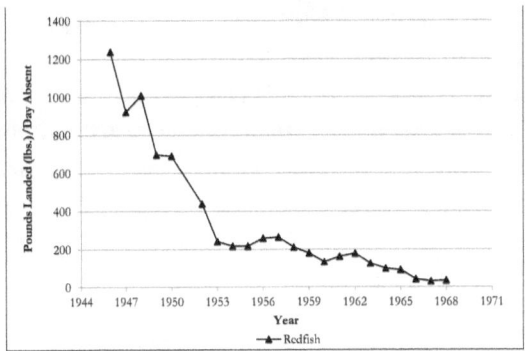

FIGURE 8: Redfish LPDA, New England waters (ICNAF Area XXII), 1946–1968. Data from: Annual reports in the U.S. Department of the Interior, Fish and Wildlife Service, "Landings at Certain Massachusetts Ports," in *Fishery Statistics of the United States*.

New England fishermen going after Nova Scotian redfish and haddock stocks saw similar patterns develop (fig. 9). On the Scotian Shelf, redfish landing rates peaked at roughly 8,800 pounds (4.0 mt) per day in 1949, fell to roughly 3,600 pounds (1.6 mt) per day by 1953, recovered to roughly 8,000 pounds (3.6 mt) in 1958, and then imploded to a little over 600 pounds (0.3 mt) by 1968.

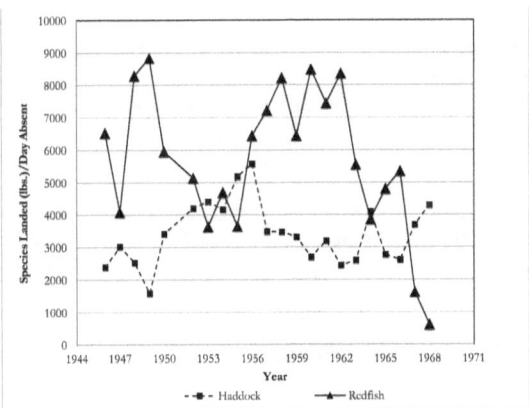

FIGURE 9. Haddock and redfish LPDA, Nova Scotia waters (ICNAF Area XXI), 1946–1968. Data from: Annual reports in the U.S. Department of the Interior, Fish and Wildlife Service, "Landings at Certain Massachusetts Ports," in *Fishery Statistics of the United States*.

Like New England stocks, Nova Scotia haddock appeared to hold its own. LPDA rates for Nova Scotia haddock actually rose fairly steadily until 1956 (peaking at 5,500 pounds [2.5 mt]) per day, sagged to 2,400 pounds (1.1 mt) per day by 1962, then rose modestly until 1968 (fig. 10). Again, surface impressions masked ominous trends. Even clearer than in New England, undersized scrod haddock comprised a steadily increasing proportion of the total catch: from a low of 5.6 percent in 1946 to a high of 75.2 percent in 1967.

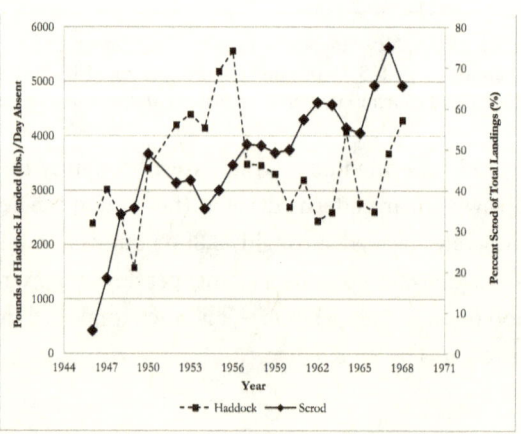

FIGURE 10: Haddock LPDA and percent scrod of total landings, Nova Scotia waters (ICNAF Area XXI), 1946–1968. Data from: Annual reports in the U.S. Department of the Interior, Fish and Wildlife Service, "Landings at Certain Massachusetts Ports," in *Fishery Statistics of the United States*.

Nova Scotia was only an option for New England vessels large enough to profitably make the trip, however. For the small and medium otter trawlers with smaller operational ranges, there was little escape from the effects of overfished local stocks.

These figures reveal a clear pattern of repeated expansion and collapse. By the early 1960s in New England and Nova Scotia, neither fishermen nor fish could sustain the intensive industrial fishery that had emerged forty years earlier. Between 1925 and 1965, otter trawling had not only depressed healthy, lightly fished standing stocks of haddock; it had also fished to near commercial extinction previously unfished stocks of redfish. Cod escaped oblivion for a time because it was harder to process in mechanized frozen fillet plants. Its day came later, after 1976. Otter trawling strained the region's marine ecosystems, but the collapse of these commercial stocks sent shockwaves throughout the region's fishing economy, spilling onto shore in the forms of labor unrest, lockouts, plummeting profitability, and more demands for federal subsidies. In four decades, New England's otter trawling fleet managed to fish a good portion of itself out of existence.

Romanticism, Tradition, and Reinventing History

In response to the periodic collapse of their fisheries, New England leaders from Boston and then Gloucester focused their attention on political solutions. Once Soviet Bloc factory trawlers arrived, they offered a convenient scapegoat for the degradation of regional fish stocks that had been underway since the 1920s. The challenge before New England's fishermen, however, was not political but ecological. Local haddock and redfish fisheries—across all vessel size classes—had simply taken too much out of the ecosystem too fast and could no longer sustain the high operational volumes needed to remain profitable. Rice and the heavy trawler leadership understood this in 1954 when they requested that Congress fund their sweeping research program on everything from the effect of otter trawls on benthic habitat to exploring other means to enhance ecological productivity. The heavy trawler fleet—labor and management alike—acknowledged its industry's ecological problems. Politicians, enamored with caricatures of New England's fishing heritage, did not. As fishermen drove down regional stocks of haddock and redfish, politicians enhanced their ability to do so—despite evidence and warnings of the consequences.

In public comments before Congress and in the press, these fundamental trends in biological productivity, which contemporary scientists had identified, were marginalized in discussion or silenced altogether. That tacit censorship continued in the 1960s and 1970s even as resources further declined. A good example of this collective myopia was Frank Bell's report *The Economics of the New England Fishing Industry: The Role of Technological Change and Government Aid*. This 1966 report ignored the productivity of local stocks and focused exclusively on gear and other costs in New England fishing. While a wholesale biological inquiry lay outside the scope of Bell's study, earlier economic analyses, such as that by Edward Lynch and colleagues in 1961, considered the fundamental health of the stock to be a critical economic factor that would determine the future of the industry. Even when Bell's research begged for consideration of the question—for example, when captains surveyed indicated that on over two-thirds of the their trips, they failed to fill their holds before returning to port—Bell did not address biological sustainability.[6] By 1966, however, it might have been pointless. By the mid-1960s and early 1970s, New Englanders took it on faith that foreign fishing vessels had destroyed New England's groundfishery.

In this political context, Boston's haddock fleet and its political leadership represented an historical inconvenience. The existence and disappearance of such a large-scale operation challenged the accepted narrative that everything was fine before the arrival of foreigners in 1961. It made far more sense for the new leadership from Gloucester to ignore Boston's fleet and its failed campaigns, and present itself as the new face of New England's fishing tradition and heritage. With Boston's large vessels—once the pride of the industry—effectively erased from public view, Gloucester's victimization- and heritage-based legislative strategy carried far more weight. Who could ascribe significant declines in catch efficiency to the small operations that now embodied America fishing traditions? Who would not stand with the small-vessel operators challenging the size, might, and military efficiency of the Soviet fishing fleets (and those of U.S. allies)? And once foreign fleets were expelled, who would not agree that Gloucester's version of the New England fishery provided the right stewardship for local marine resources? Without Boston's heavy trawler fishery taking the lead both publicly and politically, it was easy to see fishing not as an industry but as the artisanal manifestation of New England history, culture, and values that many people, both within and without the industry, had a hand in making.

Into this milieu Gloucester emerged as the region's leader in fisheries politics. Stepping into the void left by the demise of the heavy trawler industry, and joining forces with other fishing interests across the nation, Gloucester mounted new calls for industry support in the late 1960s for vessel construction subsidies, and again in the 1970s for the cessation of Georges Bank oil exploration. In the early 1970s, however, Gloucester's leadership rose to national prominence as the country considered how best to manage its severely depleted fish stocks. Working with Massachusetts representative Gerry Studds, Gloucester leaders championed what would become the 1976 Magnuson-Stevens Fisheries Conservation and Management Act (MSA), which not only pushed foreign fleets two hundred miles off the American coastline—and off the continental shelf where most fishing took place—but also gave fishing industry representatives from across the nation unprecedented influence over the future management of marine resources.[7]

As fishing communities from across the nation campaigned in Washington and in local media to win support from their congressional delegations, New England fishing communities returned to old arguments that had worked to varying degrees since the 1860s. Led by Gloucester but joined by many other smaller ports like New Bedford and others in Maine, fisheries leadership argued that New England's "traditional" small-scale fishing operations, the personification of regional and national history and culture, were better stewards for the region's marine resources than Washington bureaucrats who had allowed foreign fleets to pillage the region's fish stocks. Backed by *Boston Globe* coverage of the size and efficiency of foreign fishing fleets operating within sight of New England's shores, and of the plight of New England's fading groundfish fishery, the campaign proved to be compelling—especially where, for at least a century and half, people had supported the fishing industry's claims to represent tradition and heritage.[8]

The success of that romantic argument—that traditional fisheries would make better resource stewards than government agencies—brought into existence a problematic tension. In political campaigns since World War II, New England fisheries leadership had suppressed information about the biological trends observed in fish stocks and avoided acknowledging that fishing was—and always had been—a modern, adaptive, innovative, and responsive business. While New England saw fishermen as continuing ties to past heroism and regional culture, the passage of MSA tasked fishermen with the responsibility of resource as well as business stewardship.

MSA's structure reified in unsustainable ways the tensions between public service and private profit, at least for New England. First, MSA granted industry extensive data confidentiality protections. Second and more importantly, MSA devolved the development of federal fisheries regulations to eight regional councils populated by two groups: representatives of state marine fisheries directors and "interested members of the public."[9] Often, this latter category was interpreted to mean industry members, with the occasional outsider representing the public at large. This structure intended to put industry and public interest on relatively even footings in regulatory deliberations. In New England, however, where fishing played such an important cultural, social, historical, and political—as well as economic—role in regional identity, such constructive tensions never took hold. As councils assumed leadership over federally managed resources in 1977, few on the council were willing to limit the opportunities for New England's beleaguered, embattled, and culturally and politically resonant industry. A fully modern extractive industry, New England fisheries leadership continued to benefit from the sympathy of fellow New Englanders.

Culturally and politically, New England's fishermen could not simultaneously serve two masters. Over the next forty years, this tension between New England fishing as living history and New England fishing as a modern industry would lead to even heavier overfishing on some groundfish stocks. It also led to distorted public record-keeping as the region accepted on faith that fishermen had the long-term interests of the public's marine resources at heart. Ultimately, a lawsuit filed by the Conservation Law Foundation (CLF) in the mid-1990s found the New England Fishery Management Council guilty of insufficient public record keeping and failure to meet the management standards laid out in MSA. The foundational contradiction at the root of CLF's case—that an industry facing tremendous economic challenges would and could forego short-term economic survival for long-term ecological benefit—stemmed directly from MSA's reification of a romantic notion that entirely ignored the fate of New England's heavy trawler fishery. New England's fisheries were businesses first and foremost, and when push came to shove, business took top billing.

Cultural love for the ideal of New England fisheries—artisanal, traditional, and sustainable—has allowed these abuses to continue well into the twenty-first century. Groundfish management in New England has proven to be, quite literally, a serial disaster, if the National Marine Fisheries Service's declaration of groundfish disasters in 1994, 1995, and 2012 (totaling $146 million in direct federal relief funding to the industry) can be taken at face value.[10] Responses to such declarations—often in the forms of various fishing restrictions—all focus on what constitutes a "rebuilt" fishery. And for those discussions, historical baselines are defined in remarkably contemporary terms. Nowhere in current groundfish rebuilding policies does there appear any reference to the abundant fishery that, while ultimately unsustainable, once supported tens of thousands of New England workers in the twentieth-century haddock fishery. Nowhere in current discussions does there appear any mention of how degraded New England stocks had already become when MSA in essence placed industry in charge of management. These perspectives remain intentionally absent from current discussions. Willful denial of history has become a foundation of contemporary fisheries management.

In 1939, the Reynold's Printing Company published a short pamphlet, *Down to the Sea for Fish: The New Bedford Fishing Fleet*, celebrating the port's new fishing prosperity. After a story titled "The Port of New Bedford: Now One of the Leading Atlantic Coast Fishing Centers," other stories described Georges Bank, regaled readers with a reprinted history of New England's early trawl fishery from the 1928 *Atlantic Fisherman*, and detailed the port's fleet. But the booster piece ended on a curious note. The editors included an essay by a self-described "landsman," the magazine and newspaper writer Joseph Chase Allen. In it, Allen wrote, "It might be argued, that no man is better qualified to talk or write about the future of fishing than a fisherman himself and this would undoubtedly be the truth regarding any other industry." But in researching the future of New England's fishing, Allen had found that fishermen were almost unanimously pessimistic, and it was "seldom that any two will agree on theories of [sic] facts." Thus, Allen took "sights on the industry from a distance," to offer readers an unusual warning: "Man will eventually wind up the history of fishing in the same manner as that in which he began, with a hook and line. This, for the simple reason that there will be neither the object nor profit in the employment of more elaborate gear." Like fish, fears of dwindling fish stocks had come and gone over New England's fishing history. Allen wrote,

> But today, the picture is different. It is twenty years since a fish-trap really made a profit in Vineyard Sound. It is long since hand-line fisherman could make a living with his small boat. . . . Otter trawls receive the blame for much of this change, whether rightly or wrongfully. Even the fishermen themselves, generally agree that the use of otter-trawls has ruined the New England fisheries. For they are ruined, insofar as the individual fisherman is concerned. Once a man could secure a boat and gear, and with them, make a livelihood. That is no longer true. The big vessels undoubtedly make money, but their crews no longer share the profits of the trips.[11]

Allen noted that larger vessels were making money but pushing further offshore than ever before, after otter trawls had swept local grounds clean. Despite the obvious destruction, the thousands of people employed by otter trawl fishing compelled Allen to accept that the gear would remain in use: "If, in some manner, a return to older methods could be made, then improvement should certainly follow providing always, that the sins of the otter-trawl, are deserved." That's not the way of the world, Allen opined: "Progress does not travel in reverse. . . . Rather, it travels to certain objectives, over-riding any obstruction of objectors. Thus the conclusion seems plain enough, as before mentioned. The ending of New England's commercial fisheries altogether. The voyaging into ocean water for the entire supply, and the eventual wiping out of that to an extent that will make fish a luxury. Less than twenty-five years have sufficed to ruin the coastal waters for fishing; how long will it take to ruin the ocean fisheries?"[12]

Some would argue we have arrived at that point today. Within fifty years of Allen's essay, Boston's heavy trawlers had ceased to exist. Writing today, eighty years later, medium and small otter trawlers are struggling to survive more so than ever. Many claim that we arrived at this point imperceptibly. Daniel Pauly coined a technical term, "shifting baselines," to describe the tendency of contemporary fisheries observers to define healthy conditions in human generational terms—that what an observer had seen as a child was healthy, regardless of longer-term trends.[13] Allen would have offered another term, however: historical ignorance. Whether through forgetting, willful erasure, romanticizing, or mythologizing—all of which, this book has shown, played a role in shaping the representation and erasure of New England's heavy fishery past—each of these processes replaced or removed empirical realities in favor of desirable inventions.

Before we can ever hope to rebuild New England's shattered and crippled groundfish fishery, we need to see its full, empirically established history. We need to understand how well-intentioned desires for a compelling and lyrical history led to unintentional consequences that undermined the regional health of fish stocks and erased from collective memory the massive industry that relied on those resources. Ultimately, we need to confront the fact that for a century and a half, New England has relied more on lyrical myth than empirical history in establishing contemporary management structures. Instead of asking "how did we get to such a sad state of affairs?" we should be asking "how could we expect to be anywhere else?" Then, and only then, will we understand how much work lies before us to rebuild even a fraction of what we know—empirically—was once there.

NOTES

INTRODUCTION

1. W. Jeffrey Bolster, *The Mortal Sea: Fishing the Atlantic in the Age of Sail* (Cambridge, Mass.: Harvard University Press, 2012).
2. Margaret Dewar, *Industry in Trouble: The Federal Government and the New England Fisheries* (Philadelphia: Temple University Press, 1983); Peter B. Doeringer, Philip I. Moss, and David G. Terkla, *The New England Fishing Economy: Jobs, Income, and Kinship* (Amherst: University of Massachusetts Press, 1986).
3. Mark Kurlansky, *Cod: A Biography of the Fish That Changed the World* (New York: Penguin Books, 1997), and *The Last Fish Tale: The Fate of the Atlantic and Survival in Gloucester, America's Oldest Fishing Port and Most Original Town* (New York: Riverhead Books, 2009); David Dobbs, *The Great Gulf: Fishermen, Scientists, and the Struggle to Revive the World's Greatest Fishery* (Washington, D.C.: Island Press, 2000); Paul Greenberg, *Four Fish: The Future of the Last Wild Food* (New York: Penguin Books, 2011); Richard Adams Carey, *Against the Tide: The Fate of the New England Fisherman* (New York: Mariner Books, 2000).
4. Ian McKay, *The Quest of the Folk: Antimodernism and Cultural Selection in Twentieth Century Nova Scotia* (Montreal: McGill-Queens University Press, 1994); Miriam Wright, *A Fishery for Modern Times: The State and the Modernization of the Newfoundland Fishery, 1934–1968* (London: Oxford University Press, 2001).
5. Roderick Nash, *Wilderness and the American Mind* (New Haven, Conn.: Yale University Press, 1967).

CHAPTER 1: MOBILIZING A TRADITION

1. Sylvanus Smith, *The Fisheries of Cape Ann* (Gloucester, Mass.: Gloucester Times, 1915), 58.
2. American Fishery Union, *Memorial* (Gloucester, Mass.: N.p., 1885), 10, 20.
3. J. W. Collins, "The Outlook of the Fisheries," *Century Illustrated Monthly Magazine*, October 1886, 959.

4. Collins, "Outlook," 961.
5. Collins, "Outlook," 961.
6. See Howard S. Russell, *A Long Deep Furrow: Three Centuries of Farming in New England* (Hanover, N.H.: University Press of New England, 1976); Dona Brown and Stephen Nissenbaum, "Changing New England: 1865–1945," in *Picturing Old New England: Image and Memory*, ed. William H. Truettner and Roger B. Stein (New Haven, Conn.: Yale University Press, 1999), 1–14; Dona Brown, *Inventing New England: Regional Tourism in the Nineteenth Century* (Washington, D.C.: Smithsonian Institute Press, 1995); Joseph Conforti, *Imagining New England: Explorations of Regional Identity from the Pilgrims to the Mid-Twentieth Century* (Chapel Hill: University of North Carolina Press, 2001); and Blake Harrison, *View from Vermont* (Lebanon, N.H.: University Press of New England, 2006).
7. Bruce Robertson, "Perils of the Sea," in Truettner and Stein, eds., *Picturing Old New England*, 143; John Higham, *Strangers in the Land: Patterns of American Nativism, 1860–1925* (1955; New Brunswick, N.J.: Rutgers University Press, 2008), 35–67; Leonard Dinnerstein and David M. Reimers, *Ethnic Americans: A History of Immigration* (New York: Columbia University Press, 2009), 89–99.
8. Wayne O'Leary, *Maine Sea Fisheries: The Rise and Fall of a Native Industry, 1830–1890* (Boston: Northeastern University Press, 2006), 76.
9. O'Leary's title, *Maine Sea Fisheries*, underplays the in-depth analysis he presents of late nineteenth-century Massachusetts fishing fleets. His book is as much about Massachusetts as it is about Maine, and it stands as one of the best studies of this dynamic period to date.
10. For more comprehensive, scholarly work on the history of New England fisheries before the Civil War, see Lorenzo Sabine, *Report on the Principal Fisheries of the American Seas, Prepared for the Treasury Department of the United States* (Washington, D.C.: Robert Armstrong, 1853); George Brown Goode, ed., *The Fisheries and Fisheries Industries of the United States*, 7 vols. (Washington, D.C.: Government Printing Office, 1884–87); Raymond McFarland, *A History of the New England Fisheries* (New York: University of Pennsylvania, 1911); O'Leary, *Maine Sea Fisheries*; and W. Jeffrey Bolster, *The Mortal Sea: Fishing the Atlantic in the Age of Sail* (Cambridge, Mass.: Harvard University Press, 2012).
11. Smith, *Fisheries of Cape Ann*, 53. See also O'Leary, *Maine Sea Fisheries*, 48–77.
12. Data was compiled from annual vessels lists for the years 1869, 1870, 1872, 1874, 1878, 1883, 1885–87, 1889, 1891–94, 1896–98, 1900, 1903, 1904, and 1908. See *List of Vessels Belonging to the District of Gloucester* (Gloucester, Mass.: John S. E. Rogers, 1869–83), *passim*, and *List of Vessels Belonging to the District of Gloucester* (Gloucester, Mass.: Proctor Brothers, 1885–1908), *passim*.
13. *List of Vessels, September 1869*, 11.
14. Smith, *Fisheries of Cape Ann*, 69.
15. A. Howard Clarke, "The Fisheries of Massachusetts," in Goode, ed., *The Fisheries and Fisheries Industries of the United States*, section 2, *Geographical Review of the Fisheries Industries and Fishing Communities for the Year 1880* (Washington, D.C.: Government Printing Office), 146.
16. *List of Vessels, 1869–1883*; *List of Vessels, 1885–1908*.
17. *List of Vessels, 1869–1883*; *List of Vessels, 1885–1908*.
18. For a discussion of Canadians in the Gloucester fisheries, see Brian Payne, *Fishing a Borderless Sea* (Lansing: Michigan State University Press, 2010), 91–92.

19. *Boston Globe,* March 7, 1886.
20. "Knights: Sworn to Uphold Labor's Cause," *Boston Globe,* March 7, 1886.
21. W. A. Wilcox, "New England Fisheries," *Forest and Stream,* January 19, 1882, 492.
22. *Kansas City Evening Star,* January 22, 1885; J. W. Collins, *Fearful Experience of a Gloucester Halibut Fisherman, Astray in a Dory off the Newfoundland Coast in Mid-Winter* (Gloucester, Mass.: Cape Ann Breeze Press, [ca. 1883–89]).
23. It is possible that Homer had heard of Blackburn's exploits before Collins's published account. Upon his return from England, the painter moved to Prouts Neck, Maine, where the local fishing community would have likely picked up the story.
24. Paul Raymond Provost, "Winslow Homer's 'The Fog Warning': The Fisherman as Heroic Character," *American Art Journal* 22 (1990): 20–27; Roger B. Stein, *Seascape and the American Imagination* (New York: Clarkson Potter, 1975); John Wilmerding, *American Marine Painting* (New York: Harry N. Adams, 1987).
25. Nicolai Cikovsky, Jr., *Winslow Homer* (New York: Harry N. Abrams, 1990), 15–27.
26. See Brown and Nissenbaum, "Changing New England," 1–14; Brown, *Inventing New England;* Conforti, *Imagining New England;* and Harrison, *View from Vermont.*
27. Robertson, "Perils of the Sea," 143; Higham, *Strangers in the Land,* 35–67; Dinnerstein and Reimers, *Ethnic Americans,* 89–99.
28. Glenn M. Grasso, "Escaping the Maritime Revival Viewpoint," in *Fluid Frontiers: New Currents in Marine Environmental History,* ed. John Gillis and Franzisk Thorma (Cambridge, U.K.: White Horse Press, 2015), 39. For a more complete discussion of the Maritime Revival, see Glenn M. Grasso, "The Maritime Revival: Antimodernity, Class, and Culture, 1870–1940" (Ph.D. diss., University of New Hampshire, 2009).
29. Ian McKay, *The Quest of the Folk: Antimodernism and Cultural Selection in Twentieth-Century Nova Scotia* (Montreal: McGill-Queens University Press, 1994).
30. Robertson, "Perils of the Sea," 143.
31. Robertson, "Perils of the Sea," 143.
32. Ralph H. Gabriel, "Geographic Influences in the Development of the Menhaden Fishery on the Eastern Coast of the United States," *Geographical Review* 10 (1920): 91–100.
33. Barbra Garrity-Blake, *The Fish Factory: Work and Meaning for Black and White Fishermen of the American Menhaden Industry* (Knoxville: University of Tennessee Press, 1994).
34. George Brown Goode and A. Howard Clarke, "The Menhaden Fishery," in Goode, ed., *The Fisheries and Fisheries Industries of the United States,* section 5, *History and Methods of the Fisheries,* vol. 1 (Washington, D.C.: Government Printing Office), 327–415; "Fishculture: Menhaden Oil and Guano Association," *Forest and Stream,* January 17, 1889, 521.
35. *Commonwealth of Massachusetts v. Manchester,* 152 Mass., 230; *Manchester v. Massachusetts,* 139 U.S., 240.
36. Charles F. Chamberlayne, "Nationalism in State Fisheries," *Transactions of the American Fisheries Society* 21 (January 1892): 187–96, 192 (quotation). Note the backdating of the publication. Despite the appearance that Chamberlayne addressed the American Fisheries Society before he addressed Congress, it is clear from his speech that the fate of the Lapham bill had been determined before he presented this paper.
37. Chamberlayne, "Nationalism," 195.
38. *Report of the Commissioners of Fisheries and Game of the State of Maine for the Years 1891–1892* (Augusta, Maine: Burleigh and Flint, 1892), 32–33.

39. Maine Commission of Sea and Shore Fisheries, *A Memorial Relating to the Destruction of State Fisheries, Presented to Congress, March 9, 1892* (Washington, D.C.: Rufus Darby, 1892), 1–2.
40. Maine Commission, *A Memorial*, 2.
41. *Report of the Commissioners*, 33–34. See also Chamberlayne, "Nationalism," 195–96.
42. Gould detailed his experiences fighting the Lapham bill in the 1892 annual *Report of the Commissioners*. Chamberlayne recounted his experiences in "Nationalism."

CHAPTER 2: THE BENEFITS OF MODERN FISHING

1. Andrew W. German, *Down on T Wharf: The Boston Fisheries as Seen Through the Photographs of Henry D. Fisher* (Mystic, Conn.: Mystic Seaport Museum, 1982), 3.
2. W. A. Wilcox, "New England Fisheries," *Forest and Stream*, January 19, 1882, 492. On the changing location of Boston's fish markets, see German, *Down on T Wharf*, 3.
3. *Boston Globe*, November 13, 1888.
4. German, *Down on T Wharf*, 3.
5. "Not a Private Affair: Boston's Big Fish Trust Is in Deep Water," *Boston Globe*, November 10, 1888; "Breaks Up in a Row," *New York Times*, November 13, 1888.
6. "A Halibut Trust, Too," *New York Times*, November 15, 1888.
7. "Elephant on Their Hands: Gigantic Fish Trust Bids Fair to Founder," *Boston Globe*, February 5, 1891.
8. "Fishy Flavor: Observable in the Gossip in the Street. Talked-Of Codfish Trust Has Caught the Brokers," *Boston Globe*, April 7, 1894, 7.
9. Massachusetts General Court, *Report of the Joint Special Committee on Investigation of the Fish Industry* (Boston: Wright and Potter, 1918), 1–36.
10. "Fishing Combine Plan to Center All Interests at New York a Railroad on Long Island Part of the Scheme," *Boston Journal*, December 19, 1895.
11. "Everything Points to Success," *Boston Globe*, January 4, 1898; "Big Fish Pool," *Boston Globe*, March 24, 1899; "No Fish Trust," *Boston Globe*, July 21, 1899; "Salt Fish Trust No[t] Probable," *Boston Globe*, July 28, 1901.
12. Frederick F. Dimick, *Boston Fish Bureau Annual Report [for] January, 1892* ([Boston]: N.p., 1892), 7.
13. Leonee Ormond, introduction, in Rudyard Kipling, *Captains Courageous* (Oxford: Oxford University Press, 1995), xvii–xxi.
14. "Boston Amused," *Boston Globe*, December 19, 1895; "Organization of Sailormen," *Boston Globe*, December 28, 1902; "Stupid and Suicidal," *Boston Journal*, February 18, 1896; "No Use for It," *Boston Journal*, August 29, 1898; "How New England Was Threatened," *Boston Journal*, September 27, 1904.
15. *Boston Globe* (Extra), November 15, 1899, 1.
16. "Not One Man Lost" and "Remarkable Immunity of Provincetown Fishermen from Disaster during Past 12 Months," *Boston Globe*, November 2, 1899.
17. "Water Front Items: Fishing Schooner Susan R. Stone Probably Lost," *Boston Globe*, December 19, 1897; "Were Adrift for Twenty-Four Hours," *Boston Globe*, January 12, 1899; "Not One Man Lost," *Boston Globe*, November 2, 1899; "Heroic Act," *Boston Globe*, November 15, 1899; "Fear That They Are Lost," *Boston Globe*, March 6, 1900; "Given up for Lost," *Boston Globe*, March 7, 1900; "Provincetown Man Lost," *Boston Globe*, June 30, 1900; "Lying Dead on the Beach," *Boston Globe*, November 24, 1902;

"Save from Shovelfull," *Boston Globe*, November 25, 1902; "Boat a Wreck," *Boston Globe*, December 16, 1903; "Will Look for Bodies," *Boston Globe*, January 18, 1904; "Cling to Dory," *Boston Globe*, May 4, 1904; "The Deep Sea Fishermen," *Boston Journal*, March 12, 1897.
18. "Seven Days on the Banks," *Boston Globe*, January 24, 1891.
19. "Trawling," *Boston Globe*, December 31, 1899.
20. "Regular Tar Every Inch of Him," *Boston Globe*, June 25, 1901.
21. "Maine's Fisherman-Preacher," *Boston Globe*, September 4, 1904.
22. Francis Rolt-Wheeler, *The Boy with the US Fisheries* (Boston: Lothrop, Lee, and Shepard, 1912).
23. Ormond, introduction, xvii–xxi.
24. Rudyard Kipling, *Something of Myself* (London: Macmillan, 1937), 130, as cited in Ormond, introduction, xv.
25. See Daniel Karlin, "Captains Courageous and American Empire," *Kipling Journal* 63 (1989): 11–21.
26. Leonee Ormond, "Note on the Text," in Kipling, *Captains Courageous*, xxxiii.
27. "'Captains Courageous'—A Review," *Atlantic Monthly*, December 1897.
28. Roger Lancelyn Green, *Kipling: The Critical Heritage* (New York: Barnes and Noble, 1971), 225, 227, 275, 338, 386.
29. Daniel Vickers, "The First Whalemen of Nantucket," *William and Mary Quarterly*, 3rd series, 40 (1983): 560–83; Brian Payne, *Fishing a Borderless Sea* (Lansing: Michigan State University Press, 2010), 1–28; Ian McKay, *The Quest of the Folk: Antimodernism and Cultural Selection in Twentieth-Century Nova Scotia* (Montreal: McGill-Queens University Press, 1994), 116–19.
30. John Higham, *Strangers in the Land: Patterns of American Nativism, 1860–1925* (1955; New Brunswick, N.J.: Rutgers University Press, 2008), 158–93.
31. Winfield Thompson, "The Passing of the New England Fisherman," *New England Magazine*, February 1896, 675, 676–77, 680, 679.
32. Thompson, "The Passing of the New England Fisherman," 676–77, 679.
33. "Gigantic Fish Trust Forms," *Boston Globe*, November 15, 1906.
34. "Boston's Great Fish Industries to Be Absorbed into One Combination," *Boston Journal*, November 15, 1906.
35. Robb Robinson, *Trawling: The Rise and Fall of the British Trawl Fishery* (Exeter, England: University of Exeter Press, 1996).
36. "Beam Trawling," *Boston Globe*, April 10, 1891.
37. J. W. Collins, "The Beam Trawl Fishery of Great Britain," *Bulletin of the United States Fish Commission for 1887* 7 (1889): 289–407.
38. Dimick, *Boston Fish Bureau*, 46–48.
39. "Yankee Beam-Trawling," *Boston Globe*, December 14, 1891.
40. "Gigantic Fish Trust Forms," *Boston Globe*, November 15, 1906.
41. "Steam Trawls Spell Doom for Fishermen," *Boston Journal*, June 28, 1905.
42. "Gigantic Fish Trust Forms," *Boston Globe*, November 15, 1906.
43. "Gigantic Fish Trust Forms."
44. "Captains Prefer to Wait," *Boston Globe*, November 22 1906.
45. *List of Vessels Belonging to the District of Gloucester* (Gloucester, Mass.: John S. E. Rogers, 1908), 25.
46. B. A. Balcom, "Technology Rejected: Steam Trawlers and Nova Scotia, 1877–1933," in *How Deep Is the Ocean: Historical Essays on Canada's Atlantic Fishery*, ed. James E.

Candow and Carol Corbin (Louisbourg, Nova Scotia: University College of Cape Breton Press, 1997), 161–74.
47. Press release reprinted in James B. Connolly, *The Deep Sea's Toll* (New York: Scribner's Sons, 1905), 317.
48. House Committee on Merchant Marine and Fisheries, *Hearings before the Committee on Merchant Marine and Fisheries . . . on H.R. 16457 Prohibiting the Importing and Landing of Fish Caught by Beam Trawlers* (Washington, D.C.: Government Printing Office, 1913), 76–86 (hereafter *Beam Trawlers*).
49. *Beam Trawlers*, 83–86.
50. House Committee on Merchant Marine and Fisheries, *Otter-Trawling: Letter from the Secretary of Commerce Transmitting from the Commissioner of Fisheries Submitting a Report on the Otter-Trawl Fishery . . . , 63rd Congress, 3rd session* (Washington, D.C.: Government Printing Office, 1915).
51. In the court's final findings, the BFMC was found to include the New England Fish Exchange, to which all wholesale fresh fish dealers belonged; the Bay State Fishing Company, which owned *Spray* and several other steam trawlers by 1918; the Boston Fish Pier Company; the Commonwealth Ice and Cold Storage Company; R. O'Brien & Company; Booth Fisheries; Bunting and Emery; and the Gloucester Fresh Fish Company. See *Boston Globe*, July 12, 1919; *United States v. New England Fish Exchange et al.*, 258 F.732 (U.S. Dist. 1919); *Bay State Fishing Company v. United States*, 57 Ct. Cl. 64 (U.S. Ct. Cl. 1922); *Commonwealth v. Dyer et al.*, 243 Mass. 472, 138 N.E. 296 (Mass. 1923); and *United States v. New England Fish Exchange et al.*, 292 F. 511 (U.S. Dist. 1923).
52. Massachusetts General Court, *Report of the Joint Special Committee on Investigation of the Fish Industry*, 18–30, 117. Prices listed are presumed to be for hundredweights, no units listed in original.
53. For a detailed narrative of the formation of the Boston Fish Market Corporation, and the various other entities named in the Fish Trust case, see Massachusetts General Court, *Report of the Joint Special Committee on Investigation of the Fish Industry*, 11–29.
54. "Government Opens War on Boston's 'Fish Trust,'" *Boston Globe*, June 22, 1917; "Boston Ex-Mayor Attacks Fish Trust," *Boston Globe*, August 2, 1918; "Indict 30 on Fish Monopoly Charge," *Boston Globe*, August 16, 1918.
55. Massachusetts General Court, *Report of the Joint Special Committee on Investigation of the Fish Industry*, 30.
56. Massachusetts General Court, *Report of the Joint Special Committee on Investigation of the Fish Industry*, 34.
57. Commonwealth of Massachusetts, *Report of the Joint Special Recess Committee to Continue the Investigations of the Fish Industry* (Boston: Wright and Potter, 1919), 36.
58. Commonwealth of Massachusetts, *Report of the Joint Special Recess Committee*, 35–39.
59. Commonwealth of Massachusetts, *Report of the Joint Special Recess Committee*, 63–64.
60. The court cases ensuing from the Supreme Judicial Court's investigations targeted not only the BFMC but also many of its myriad subsidiaries. See *United States v. New England Fish Exchange et al.*, 258 F.732 (U.S. Dist. 1919); *Bay State Fishing Company v. United States*, 57 Ct. Cl. 64 (U.S. Ct. Cl. 1922); *Commonwealth v. Dyer et al.*, 243 Mass. 472, 138 N.E. 296 (Mass. 1923); and *United States v. New England Fish Exchange et al.*, 292 F. 511 (U.S. Dist. 1923).

61. "Fish Trust Men Have Best Island Affords," *Boston Globe*, April 7, 1923.

CHAPTER 3: MASKING INDUSTRIAL REALITIES

1. Wayne Santos, *Caught in Irons: North Atlantic Fishermen in the Last Days of Sail* (Selinsgrove, Penn.: Susquehanna University Press, 2002), 84–99. Santos offers one of the most cogent analyses of these races, highlighting the interplay of changing workplace culture, romanticism, tourism, and boosterism. I am heavily indebted to Professor Santos for his ideas and insights.
2. Santos, *Caught in Irons*, 23–32.
3. Santos, *Caught in Irons*, 84–86.
4. Frederick Wallace, "Life on the Grand Banks," *National Geographic*, July 1921, 4, 5.
5. Wallace, "Life on the Grand Banks," 4, 5.
6. Wallace, "Life on the Grand Banks," 28.
7. *Report of the United States Commissioner of Fisheries for the Fiscal Year 1922* (Washington, D.C.: Government Printing Office, 1923), 51–53.
8. *Report of the United States Commissioner of Fisheries for the Fiscal Year 1924* (Washington, D.C.: Government Printing Office, 1925), 178, 181.
9. "Investing in Steam Trawlers: The Much Discussed Vessel Has Been a Bitter Disappointment as a Profit Producer," *United States Investor* 31 (December 11, 1920): 3039.
10. "Gorton-Pew Fisheries," *United States Investor* 31 (July 10, 1920): 1604.
11. "The Crash in the East Coast Fisheries," *United States Investor* 31 (December 18, 1920): 3101.
12. "Investing in Steam Trawlers," 3040, 3048.
13. "Investing in Steam Trawlers," 3048.
14. "From the Drought, an Opportunity," *Fortune*, April 1935, http://fortune.com.
15. "From the Drought, an Opportunity."
16. "From the Drought, an Opportunity."
17. A. W. H. Needler, "Statistics of the Haddock Fishery in North American Waters," in *Report of the United States Commissioner of Fisheries for the Fiscal Year 1930* (Washington, D.C.: Government Printing Office, 1931), 27–40.
18. Raymond C. Mudge, "Net Trawl for Ground Fishing," U.S. Patent Office, application filed October 5, 1923; patent no. 1,600,839, granted September 21, 1926.
19. *Report of the United States Commissioner of Fisheries for the Fiscal Year 1926* (Washington, D.C.: Government Printing Office, 1927), 252–55; *Report of the United States Commissioner of Fisheries for the Fiscal Year 1928* (Washington, D.C.: Government Printing Office, 1929), 462–63.
20. *Report of the United States Commissioner of Fisheries for the Fiscal Year 1928*, 462–63.
21. *Report of the United States Commissioner of Fisheries for the Fiscal Year 1929* (Washington, D.C.: Government Printing Office, 1930), 461–63.
22. *Report of the Joint Special Committee on Investigation of the Fish Industry* (Boston: Wright and Potter, 1918), 34.
23. George Brown Goode. ed., *The Fisheries and Fisheries Industries of the United States* (Washington D.C.: Government Printing Office, 1884–87).
24. Francis Parkman's monumental series *France and England in North America* (1865–92) best represents his stylistic vision of the American settler. See in particular his later volumes: part 5, *Montcalm and Wolfe* (Boston: Little, Brown, 1884), and part 6, *Half Century of Conflict* (Boston: Little, Brown, 1897). On Parkman's romanti-

cism, see Wilbur R. Jacob, "Francis Parkman's Oration 'Romance in America,'" *American Historical Review* 68 (April 1963): 692–97. For more current discussions of Parkman's style and historiography, see Wilbur R. Jacobs, "Francis Parkman—Naturalist—Environmental Savant," *Pacific Historical Review* 61 (May 1992): 341–56; Stephen P. Knadler, "Francis Parkman's Ethnography of the Brahmin Caste and *The History of the Conspiracy of Pontiac*," *American Literature* 65 (June 1993): 215–38; Wilbur R. Jacobs, "Francis Parkman and Frederick Jackson Turner Remembered," *Proceedings of the Massachusetts Historical Society*, 3rd ser., 105 (1993): 39–58; and Albert Hurtado, "Parkmanizing the Spanish Borderlands: Bolton, Turner, and the Historians' World," *Western Historical Quarterly* 26 (Summer 1995): 149–67.

25. Raymond McFarland, *A History of the New England Fisheries* (New York: D. Appleton, 1911), vi.
26. Carl Russell Fish, review of *A History of the New England Fisheries*, by Raymond McFarland, *American Economic Review* 1 (September 1911): 572.
27. James Scott Brown, review of *A History of the New England Fisheries*, by Raymond McFarland, *American Journal of International Law* 5 (July 1911): 865–66.
28. See chapters 5 and 6, below.
29. Samuel Eliot Morison, *The Maritime History of Massachusetts* (Boston: Houghton Mifflin, 1921), 137–38.
30. Morison, *Maritime History of Massachusetts*, 135, 106; "Fishermen Vote to Strike on July 3," *Boston Globe*, June 21, 1919; "Fish Controversy Near Settlement," *Boston Globe*, November 7, 1919; Santos, *Caught in Irons*, 66.
31. Santos, *Caught in Irons*, 99–112.
32. "Clarence Birdseye," in *American National Biography*, ed. John A. Garraty (New York: Oxford University Press and American Council of Learned Societies, 1999), 2:808–9.
33. *The Rudder*, April 1930, cited in Santos, *Caught in Irons*, 139. See also Santos, *Caught in Irons*, 137–39.
34. Santos, *Caught in Irons*, 142–60.
35. *Report of the United States Commissioner of Fisheries for the Fiscal Year 1925* (Washington, D.C.: Government Printing Office, 1926), 266.
36. Henry B. Bigelow, "A Developing Viewpoint in Oceanography," *Science*, January 24, 1930, 84–89; Henry B. Bigelow, *Oceanography: Its Scope, Problems, and Economic Importance* (Boston: Houghton Mifflin, 1931).
37. For a historical overview of the European international development of fisheries science and a history of ICES, see Helen Rozwadowski, *The Sea Knows No Boundaries: A Century of Marine Sciences under ICES* (Seattle: University of Washington Press, 2002). For a scientist's perspective on the development of fisheries science during the same period, see Tim D. Smith, *Scaling Fisheries: The Science of Measuring the Effects of Fishing, 1855–1955* (Cambridge: Cambridge University Press, 1994). On Huntsman and the acceptance of the theory of overfishing, see Jennifer Hubbard, *A Science on the Scales: The Rise of the Canadian Atlantic Fisheries Biology, 1898–1939* (Toronto: University of Toronto Press, 2006), especially chapter 6.
38. William C. Herrington, "Conservation of Immature Fish in Otter Trawling," *Transactions of the American Fisheries Society* 62, no. 1 (1932): 57–58.
39. Herrington, "Conservation of Immature Fish," 58. Current research shows that Herrington was wrong in this assessment. Otter trawling has been documented as one of the most destructive forms of fishing on benthic habitat, flora, and fauna,

and habitat destruction has been shown to be a significant impediment to rebuilding New England's marine fish stocks. On the effect of bottom trawling on sea floor and the ecological impacts of such habitat destruction, see, for example, Les Watling and Elliot A. Norse, "Disturbance of the Seabed by Mobile Fishing Gear: A Comparison to Forest Clearcutting," *Conservation Biology* 12 (December 1998): 1180–97. See also Brianna K. Brown, Elizabeth Soule, and Les Kaufman, "Effects of Excluding Bottom-Disturbing Mobile Fishing Gear on Abundance and Biomass of Groundfishes in the Stellwagen Bank National Marine Sanctuary, USA," *Current Zoology* 56, no. 1 (2010): 134–43.

40. Herrington, "Conservation of Immature Fish," 59.
41. Herrington, "Conservation of Immature Fish," 59.
42. Herrington, "Conservation of Immature Fish," 59.
43. William C. Herrington and John R. Webster, "Why Are There Good and Bad Haddock Years," *Fishing Gazette* 50 (September 1933): 4.
44. Herrington and Webster, "Good and Bad Haddock Years," 5–6.
45. Herrington, "Conservation of Immature Fish," 62.
46. William C. Herrington, "Modifications in Gear to Curtail the Destruction of Undersized Fish in Otter Trawling," *United States Bureau of Fisheries Investigational Report No. 24* (Washington, D.C.: Government Printing Office, 1935).
47. *Boston Globe*, April 16, 1933.
48. House Committee on Merchant Marine, Radio, and Fisheries, *Hearings . . . on Extending Construction Loan Fund Benefits to the Fishing and Whaling Industries, U.S. House of Representatives, Seventy-Third Congress, Second Session* (Washington, D.C.: Government Printing Office, 1934) (hereafter *Fishing and Whaling Loan*).
49. "Thebaud to Carry Plea of Fishing Fleet to F.D.," *Boston Globe*, April 16, 1933.
50. "Thebaud to Carry Plea of Fishing Fleet to F.D."; "Thebaud Will Sail Tonight to Request Aid at Capitol," *Boston Globe*, April 19, 1933; "Thebaud Sails on Aid Errand," *Boston Globe*, April 20, 1933; "Sch. Thebaud Due at Washington Sunday," *Boston Globe*, April 22, 1933; "Lemonade (after Beer at Lunch) and Tour of Mansion for Them," *Boston Globe*, April 25, 1933; "Skippers Return Mrs. F. D.'s Call," *Boston Globe*, April 25, 1933.
51. "Thebaud Will Sail Tonight to Request Aid at Capitol"; "Thebaud Sails on Aid Errand."
52. "Lemonade (after Beer at Lunch) and Tour of Mansion for Them"; "Skippers Return Mrs. F. D.'s Call."
53. *Fishing and Whaling Loan*, 5.
54. *Fishing and Whaling Loan*, 8–9.
55. *Fishing and Whaling Loan*, 9.
56. *Fishing and Whaling Loan*, 18. Current scholars have also perpetuated this interpretation. See Christopher Magra, *The Fisherman's Cause: Atlantic Commerce and Maritime Dimensions of the American Revolution* (New York: Cambridge University Press, 2009).
57. *Fishing and Whaling Loan*, 20.
58. *Fishing and Whaling Loan*, 20–21.
59. House Committee on Merchant Marine, Radio, and Fisheries, *Hearings . . . on [the] Rehabilitation of the Fishing Industry, U.S. House of Representatives, Seventy-Third Congress, Second Session* (Washington, D.C.: Government Printing Office, 1934), 4, 6, 11.
60. House Committee on Merchant Marine, Radio, and Fisheries, *Hearings . . . on*

[Funding a] *Research Vessel for the Bureau of Fisheries, U.S. House of Representatives, Seventy-Third Congress, Second Session* (Washington, D.C.: Government Printing Office, 1934) (hereafter *Research Vessel*).
61. *Research Vessel*, 2, 10.
62. *Research Vessel*, 11.
63. Frank S. Nugent, "Captains Courageous," *New York Times*, May 12, 1937.
64. Edward A. Ackerman, "Depletion in New England Fisheries," *Economic Geography* 14 (July 1938): 233.
65. Ackerman, "Depletion in New England Fisheries," 233, 236.
66. John R. Arnold, "The Fishery Industry and the Fishery Codes," in *Work Materials No. 31*, ed. Office of National Recovery Administration, Division of Review (N.p.: Industry Studies Review, 1936), 33.

CHAPTER 4: REINVENTING TRADITION

1. All data for discussion of changing fleet size, effort patterns, and species landed come from R. H. Fiedler, "Fisheries of the New England States," for the years 1929 to 1939, part of the annual reports published as *Fishery Industries of the United States*, which are included in the *Report of the United States Commissioner of Fisheries* (Washington, D.C.: Government Printing Office, 1931–50) (hereafter Fiedler, "Fisheries of the New England States"). Data collected between 1940 and 1966 appears under the heading "Fisheries of the New England States" in sequentially numbered statistical digests published by the U.S. Fish and Wildlife Service under the title *Fishery Statistics of the United States* (Washington, D.C.: Government Printing Office, 1942–68).
2. Fiedler, "Fisheries of the New England States."
3. Frank H. Wood, *The Story of 40-Fathom Fish* (Boston: Bay State Fishing Company, 1931), 3–6.
4. Bernard Breedlove, "Friday's Fish," *Saturday Evening Post*, September 24, 1938, 18–19, 56.
5. Edward A. Ackerman, *New England's Fishing Industry* (Chicago: University of Chicago Press, 1941), 1–5, 71.
6. Ackerman, *New England's Fishing Industry*, 61–65, chapter 5, 80–81.
7. Ackerman, *New England's Fishing Industry*, 81.
8. Ackerman, *New England's Fishing Industry*, 123.
9. Ackerman, *New England's Fishing Industry*, 125.
10. Ackerman, *New England's Fishing Industry*, 143.
11. Ackerman, *New England's Fishing Industry*, 146.
12. Ackerman, *New England's Fishing Industry*, 123.
13. *Boston Globe*, October 2, 7, 9, 15, 20, 23, 1934; *New York Times*, October 10, 1934.
14. "Fishermen's Story Told at Food Fair," *Boston Globe*, October 20, 1934.
15. "Ten Trawlers Sail for Fishing Grounds," *Boston Globe*, October 26, 1934.
16. "Threaten Trawler Strike on March 1," *Boston Globe*, February 17, 1939.
17. "Trawler Labor Pact Ends Strike Threat," *Boston Globe*, February 28, 1939.
18. For 1940, see, in chronological order, "Union Won't Restrict Fishing Boats at Sea, Pending Negotiations," *Boston Globe*, March 17, 1940; "Union Fishermen to Consider Ban on Fish Pier," *Boston Globe*, March 20, 1940; "Asks U.S. Probe of 'Practices' at Fish Pier," *Boston Globe*, March 26, 1940. For 1942, see "Fishermen Strike

for War Aid," *New York Times*, January 9, 1942; "Board Rejects Fishing Operators' Reply," *Boston Globe*, February 1, 1942; "U.S. May Be Asked to Run Fishing Fleet," *Boston Globe*, February 13, 1942; "Fishing Boat Owners to Accept Truce Terms, to Resume Operations," *Boston Globe*, February 15, 1942; "General Seafoods Strike Hits Boston, Gloucester Piers," *Boston Globe*, August 12, 1942. For 1943, see "Fishermen May Quit," *New York Times*, July 11, 1943; "Fishermen Will Meet with WLB to Avert Tieup," *Boston Globe*, July 2, 1943; "N.E. Fish Strike Threatened in Protest of Price Ceilings," *Boston Globe*, July 15, 1943; "47 Boston Fishing Boats Tied Up," *Boston Globe*, July 17, 1943; "WLB Bans Fish 'Strike,'" *New York Times*, July 21, 1943; "Boston Fishermen Vote to End Strike," *New York Times*, July 24 1943; "1000 Fishermen Vote on Strike Question in Boston Sept. 28," *Boston Globe*, September 16, 1943; Bill Godsoe, "Waterfront Spotlight: Next Move up to Government Agencies," *Boston Globe*, October 10, 1943; "Congress Committee Told Fish Industry in Black Mart Grip," *Boston Globe*, November 18, 1943; "Union Calls Strike on Fishing Fleets," *New York Times*, November 21, 1943; "Congress Probe of Fishing Strike Ends Hopefully," *Boston Globe*, December 3, 1943; "End of Fish Strike Forecast as Crews Call Sunday Meeting," *Boston Globe*, December 31, 1943. For 1944, see Lawrence Dame, "Dispute on Prices Holds up Fishing," *New York Times*, January 2, 1944; Bill Godsoe, "Waterfront Spotlight: High Hopes for Fishing Industry," *Boston Globe*, January 23, 1944; "Fishermen Return," *Business Week*, July 22, 1944, 30–32; "Backs Fish Ceilings to Guard Crews' Pay," *New York Times*, October 24, 1944. For 1945, see "Labor Board Orders Strike and Lockout in Fish Fleet Case," *Boston Globe*, September 8, 1945; "New Fishing Tie-Up Looms as 'Medford' Crew Defies WLB," *Boston Globe*, September 13, 1945. For 1946, see "Trawler Strike Cuts Hub Fish Catch 80 Percent," *Boston Globe*, February 10, 1946; "Fishing Firms Threaten Move to Other Ports," *Boston Globe*, March 7, 1946; "Fishing Union Denies Ban on Veterans Joining," *Boston Globe*, March 28, 1946.

19. U.S. House of Representatives, Committee on Merchant Marine and Fisheries, *Hearings before the Committee on Merchant Marine and Fisheries on H.R. 8475 "A Bill to Define an American Fishery,"* Seventy-Sixth Congress, Third Session, April 16, 1940 (Washington, D.C.: Government Printing Office, 1940), 25–29, 34–35 (hereafter *Define an American Fishery*).
20. "Treasury Decision 49682, August 18, 1938," in *Define an American Fishery*, 141.
21. *Define an American Fishery*, 25–29.
22. *Define an American Fishery*, 25–29.
23. Maxwell Frederic Coplan, "Cod and Canvas," *Saturday Evening Post*, August 3, 1939, 63–64.
24. Coplan, "Cod and Canvas," 63–64.
25. Coplan, "Cod and Canvas," 63–64.
26. Rachel Carson, *Food from the Sea: Fish and Shellfish of New England*, Conservation Bulletin 33, U.S. Fish and Wildlife Service (Washington, D.C.: Government Printing Office, 1943), 46–47.
27. Fiedler, "Fisheries of the New England States," 1939, p. 283.
28. William C. Herrington, *A Crisis in the Haddock Fishery* (Washington, D.C.: Government Printing Office, 1941).
29. Fiedler, "Fisheries of the New England States," 1931–38.
30. *Down to the Sea for Fish: New Bedford Fishing Fleet* ([New Bedford, Mass.]: Reynolds, 1939), *passim*.

31. Margaret Dewar, *Industry in Trouble: The Federal Government and the New England Fisheries* (Philadelphia: Temple University Press, 1983), 14–15.
32. Carson, *Food from the Sea.*

CHAPTER 5: THE CONSEQUENCES OF MODERN FISHING

1. Gerald Roscoe, "Otter Trawls Boost Area to Top Place in Industry," *Boston Globe*, October 13, 1946; Gerald Roscoe, "Quick Freeze Speeds Food from Area to Midwest," *Boston Globe*, October 14, 1946; Gerald Roscoe, "Boston One of Greatest Fishing Ports in World," *Boston Globe*, October 16, 1946; Gerald Roscoe, "Gloucester Leads Nation in Annual Fish Catches," *Boston Globe*, October 18, 1946.
2. Gerald Roscoe, "Behind the Price Collapse: N. E. Fishing Industry Set to Battle for Life," *Boston Globe*, February 9, 1947.
3. Roscoe, "Behind the Price Collapse."
4. "Fishermen, State to Test Injunction at Pier Tomorrow," *Boston Globe*, February 23, 1947; "Fishing Union Asks Owners to Stagger Boat Schedules," *Boston Globe*, February 27, 1947; "Buyer Testifies Union Barred His High Bid for Fish," *Boston Globe*, June 5, 1947; "Fish 'Limit' Voted Here 2 Years Ago, Union Man Claims," *Boston Globe*, June 20, 1947; "Threat of 'Limit' on Fish Catches Described at Union Trial Here," *Boston Globe*, June 13, 1947.
5. Margaret Dewar, *Industry in Trouble: The Federal Government and the New England Fisheries* (Philadelphia: Temple University Press, 1983), 99.
6. "Judge Rules against Union in Fish Price Case," *Boston Globe*, August 1, 1947; "Union Ruled a Monopoly in Fish Prices Case," *Boston Globe*, August 2, 1947;
7. "Union Ruled a Monopoly, in Fish Price Case."
8. "Buyer Testifies Union Barred His High Bid for Fish."
9. "Fish 'Limit' Voted Here 2 Years Ago."
10. "Judge Rules against Union in Fish Price Case."
11. "Threat of 'Limit' on Fish Catches Described at Union Trial Here"; "Judge Rules against Union."
12. *Commonwealth of Massachusetts v. McHugh et al.*, 71 F. Supp. 516 (U.S. Dist. 1947); *Atlantic Fishermen's Union et al. v. Barnes, Attorney General of Massachusetts*, 71 F. Supp. 927 (U.S. Dist. 1947).
13. "Labor: Monopoly Broken," *Time*, September 15, 1947, http://content.time.com.
14. Milton McKaye, "Good-By to the Dory Trotters," *Saturday Evening Post*, February 8, 1946, 34, 35.
15. McKaye, "Good-By," 64, 66.
16. McKaye, "Good-By," 68.
17. McKaye, "Good-By," 66.
18. Roscoe, "Otter Trawls"; Roscoe, "Quick-Freeze"; Roscoe, "Boston One of Greatest"; Roscoe, "Gloucester Leads Nation."
19. James Higgins and Gordon Donald, "Trouble at the Fish Pier," *New Republic*, September 1, 1947, 22.
20. Higgins and Donald, "Trouble at the Fish Pier," 23.
21. Higgins and Donald, "Trouble at the Fish Pier," 23.
22. Higgins and Donald, "Trouble at the Fish Pier," 23.
23. Edmund Gilligan, "Yankee Fish Famine," *Colliers Weekly*, December 7, 1946, 22–23, 105.

24. Alfred Perlmutter, "The Blackback Flounder and Its Fishery in New England and New York," *Bulletin of the Bingham Oceanographic Collections* 11 (November 1947): 1–92. I am grateful to Mark Gibson, Rhode Island Department of Environmental Management, Fish and Wildlife Division, for identifying this research.
25. "Fishing Industry's Woes Not All Due to Foreign Imports," *Boston Globe*, April 6, 1950.
26. Donald J. White, "The New England Fishing Industry: America's Oldest Industry Faces Crisis," *Monthly Review of the Federal Reserve Bank of Boston*, March 1950, reprinted in *U.S. Senate, Hearings before a Subcommittee of the Committee on Interstate and Foreign Commerce, Eighty-First Congress, Second Session, on S. 2801, "A Bill to Give Effect to the International Convention for the Northwest Atlantic Fisheries..." April 4 and 5, 1950* (Washington, D.C.: Government Printing Office, 1950), 5 (hereafter ICNAF Hearings).
27. White, "The New England Fishing Industry," 11.
28. White, "The New England Fishing Industry," 11–12.
29. ICNAF Hearings, 20–104.
30. ICNAF Hearings, 66.
31. "10 Nations Called to Parley on Fish," *Boston Globe*, November 15, 1948.
32. ICNAF Hearings, 21.
33. ICNAF Hearings, 4.
34. White, "The New England Fishing Industry," 5.
35. ICNAF Hearings, 23, 25.
36. ICNAF Hearings, 53, 45–46.
37. ICANF Hearings, 55.
38. "A Bill to Give Effect to the International Convention for the Northwest Atlantic Fisheries, Signed at Washington under Date February 8, 1949 and for Other Purposes," in ICNAF Hearings, 1–4.
39. ICNAF Hearings, 72–73.
40. ICNAF Hearings, 74–75.
41. ICNAF Hearings, 75.
42. ICNAF Hearings, 76, 86.
43. ICNAF Hearings, 22–28, 75.
44. ICNAF Hearings, 76.

CHAPTER 6: INVOKING THE PAST, IGNORING THE PRESENT, COMPROMISING THE FUTURE

1. U.S. Merchant Marine and Shipping Act of 1920, Pub. L. no. 66–261, 41 Stat. 988 (1920).
2. Ernest Premetz, *Destruction of Undersized Haddock on Georges Bank, 1947–1951* (Washington, D.C.: Department of the Interior, 1953), 32.
3. H. A. Schuck, "Protecting Baby Scrod Raises Production," *Atlantic Fisherman*, December 1947, 2; "Destruction of Baby Haddock on Georges Bank," *Yearbook of the Fishing Masters Association* (1947): 5–7; "Current Haddock Situation on Georges Bank," *Commercial Fisheries Review* 10 (October 1951): 151–76.
4. Donald J. White, "The New England Fishing Industry: America's Oldest Industry Faces Crisis," *Monthly Review of the Federal Reserve Bank of Boston*, March 1950, reprinted in *U.S. Senate, Hearings before a Subcommittee of the Committee on Inter-

state and Foreign Commerce, Eighty-First Congress, Second Session, on S. 2801, "A Bill to Give Effect to the International Convention for the Northwest Atlantic Fisheries..." April 4 and 5, 1950 (Washington, D.C.: Government Printing Office, 1950), 11.
5. White, "New England Fishing Industry," 11–12.
6. For more on the differing labor relations within the major New England fishing ports at this time, see Donald J. White, *The New England Fishing Industry* (Cambridge, Mass.: Harvard University Press, 1954), 6–11.
7. "All Gloucester Involved in Fish Industry Strike," *Boston Globe,* June 11, 1950.
8. "All Gloucester Involved in Fish Industry Strike."
9. See *Boston Globe,* April 13, 1950; July 23, August 22, 1953; August 21, 1954.
10. Margaret E. Dewar, *Industry in Trouble: The Federal Government and the New England Fisheries* (Philadelphia: Temple University Press, 1983), 48–52.
11. Dewar, *Industry in Trouble,* 48–52.
12. Edward J. Lynch, Richard M. Doherty, and George P. Draheim, "The Groundfish Industries of New England and Canada: A Comparative Economic Analysis," U.S. Department of the Interior, Fish and Wildlife Service Circular 121 (Washington, D.C.: U.S. Fish and Wildlife Service, 1961), 2.
13. Dewar, *Industry in Trouble,* 49.
14. David Boeri, *"Tell it Good-Bye, Kiddo"* (Camden, Maine: International Marine Publishing, 1976), 87.
15. Lynch, Doherty, and Draheim, "The Groundfish Industries of New England and Canada," 104.
16. Thomas Fulham quoted in Dewar, *Industry in Trouble,* 86–87.
17. *Boston Globe,* August 10, 1951.
18. *Boston Globe,* July 3, 1954.
19. Lynch, Doherty, and Draheim, "The Groundfish Industries of New England and Canada," 39.
20. Kenneth T. Frank et al., "Trophic Cascades in a Formerly Cod-Dominated Ecosystem," *Science* 308, no. 5728 (2005): 1621–23; Robert Steneck and Enric Sala, *Large Marine Carnivores: Trophic Cascades and Top-Down Controls in Coastal Ecosystems Past and Present* (Orono: University of Maine Sea Grant College Program, 2005); Georgi M. Daskalov et al., "Trophic Cascades Triggered by Overfishing Reveal Possible Mechanisms of Ecosystem Regime Shifts," *Proceedings of the National Academy of Sciences* 104, no. 25 (2007): 10518–523.
21. U.S. House of Representatives, *Hearings before a Subcommittee on Merchant Marine and Fisheries on H.R. 4303, H. R. 5829, H. R. 7441, H. R. 7641, and H. R. 7671, To Further Encourage the Distribution of Fishing Products* (Washington, D.C.: Government Printing Office, 1954), 47–50 (hereafter Saltonstall-Kennedy House Hearings).
22. Saltonstall-Kennedy House Hearings, 45–46.
23. Saltonstall-Kennedy House Hearings, 43.
24. Saltonstall-Kennedy House Hearings, 53.
25. Saltonstall-Kennedy House Hearings, 57.
26. Saltonstall-Kennedy House Hearings, 11.
27. Saltonstall-Kennedy House Hearings, 12.
28. Saltonstall-Kennedy House Hearings, 22–24.
29. Saltonstall-Kennedy House Hearings, 29–31, 52–57.
30. Saltonstall-Kennedy Act, 68 Stat. 376; 15 U.S.C. §713c-3.

31. U.S. Senate, *Hearings before the Committee on Interstate and Foreign Commerce on S. 2379, S. 3275, S. 3339, 84th Congress, 2nd Session* (Washington, D.C.: Government Printing Office, 1956), 10, 11, 33, 58 (hereafter 1956 Senate Fisheries Hearings).
32. 1956 Senate Fisheries Hearings, 10, 11, 33, 58.
33. 1956 Senate Fisheries Hearings, 119–21.
34. 1956 Senate Fisheries Hearings, iii–v.
35. 1956 Senate Fisheries Hearings, 53–77.
36. 1956 Senate Fisheries Hearings, 268.
37. 1956 Senate Fisheries Hearings, 268, 281, 160–65.
38. U.S. Senate, "Report to Accompany S. 3275, Fisheries Act of 1956," Calendar no. 2040, Report no. 2017, 84th Congress, 2nd Session (May 17, 1956).
39. U.S. Senate, *Hearings before the Committee on Interstate and Foreign Commerce, on S. 237 . . . S. 2719 . . . S. 3229 . . . [and] S. 3530, 85th Congress, 2nd Session* (Washington, D.C.: Government Printing Office, 1958), 110–11 (hereafter 1958 Senate Fisheries Hearings).
40. 1958 Senate Fisheries Hearings, 110–11.
41. 1958 Senate Fisheries Hearings, 111.
42. 1958 Senate Fisheries Hearings, 113–14.
43. 1958 Senate Fisheries Hearings, 114, 137.
44. Based on a list of people providing statements or written comments; 1958 Senate Fisheries Hearings, vi–vii.
45. 1958 Senate Fisheries Hearings, 22–23, 140, 212.
46. 1958 Senate Fisheries Hearings, 227.
47. 1958 Senate Fisheries Hearings, 220.
48. U.S. Senate, "Report . . . to Accompany S. 3229, Fisheries Assistance Act of 1958," Calendar no. 2387, Report no. 2334, 85th Congress, 2nd Session (August 12, 1958).
49. U.S. House of Representatives, *Hearings before the Subcommittee on Merchant Marine and Fisheries on H. R. 5421, 86th Congress, 1st Session* (Washington, D.C.: Government Printing Office, 1959), 2–3 (hereafter 1959 House Fisheries Hearings).
50. 1959 House Fisheries Hearings, 2–3.
51. 1959 House Fisheries Hearings, 29–30, 31.
52. 1959 House Fisheries Hearings, 35.
53. 1959 House Fisheries Hearings, 42–43.
54. 1959 House Fisheries Hearings, 70.
55. 1959 House Fisheries Hearings, 43, 71
56. 1959 House Fisheries Hearings, 85, 135.
57. 1959 House Fisheries Hearings, 104.
58. 1958 Senate Fisheries Hearings, 100.

CONCLUSION

1. Data discussed in the following section was compiled from annual reports in the U.S. Department of the Interior, Fish and Wildlife Service, "Landings at Certain Massachusetts Ports," in *Fishery Statistics of the United States* (Washington, D.C.: Government Printing Office, 1930–68). Tables entitled "Summary of Fishery: By Gear and Subarea" and "Summary of Fishery, By Area and Subarea" present yearly data including weight of landings by species, binned by size, area fished, days at sea,

numbers of vessels, binned by size and by gear. Landings per day absent was calculated using these data. In 1962, data collectors no longer separated otter trawling vessels by size class.

2. Ernest Premetz, *U.S. Fish and Wildlife Service Special Scientific Report, Fisheries no. 96* (Washington, D.C.: U.S. Department of the Interior, Fish and Wildlife Service, 1953).

3. See Stephen Clark, William J. Overholtz, and Richard C. Hennemuth, "Review and Assessment of the Georges Bank and Gulf of Maine Haddock Fishery," *Journal of Northwest Atlantic Fishery Science* 3 (1982): 1–27.

4. Charles H. Lyles, *Fishery Statistics of the United States, 1963*, U.S. Department of the Interior, Fish and Wildlife Service Statistical Digest 57 (Washington, D.C.: Bureau of Commercial Fisheries, 1965), 131.

5. Howard A. Schuck, "Studies of Georges Bank Haddock, Part I: Landings by Pounds, Numbers, and Sizes of Fish," in *Fishery Bulletin of the Fish and Wildlife Service*, Fishery Bulletin no. 66 (Washington, D.C.: Government Printing Office, 1951), 52:152.

6. Frank Bell, *The Economics of the New England Fishing Industry: The Role of Technological Change and Government Aid*, Research Report to the Federal Reserve Bank of Boston, no. 31 (Boston: Federal Reserve Bank of Boston, 1966), 101, 113.

7. Fishery Conservation and Management Act of 1976, Public Law 94–265, 16 U.S.C. 1801–1882, 90 Stat. 331.

8. Earl Banner, "Reds Taking Everything That Swims," *Boston Globe*, August 25 1963; Earl Banner, "Fishermen Set to Buck Reds," *Boston Globe*, September 7, 1963; Charles Bartlett, "Sputnik and Fish," *Boston Globe*, September 10, 1963; William Fripp, "Trawling Is a Man's Trade," *Boston Globe*, October 1, 1967; Donald White, "U.S. Fishing Is Foundering on Tradition," *Boston Globe*, October 8, 1967; Thomas Oliphant, "For the U.S. the Future in Fish Is in the Processing," *Boston Globe*, February 25, 1968; Deckle McLean, "Fishing: In Troubled Waters," *Boston Globe*, November 9, 1969; Frank Donovan, "Depression, Despair Mark Fishing Industry," *Boston Globe*, June 27, 1970; David B. Wilson, "The New England Fishery: An Industry," *Boston Globe*, September 16, 1973.

9. The original Senate version of the bill did not provide for a New England council: New England waters fell under the jurisdiction of an Atlantic council with responsibility for an area from Diamond Shoals off Cape Hatteras to the Canadian border. The Northeast Fishery Science Center retains this jurisdictional range, providing scientific support for both the Mid-Atlantic Council, which manages from Diamond Shoals to roughly Montauk, New York, and the New England Council, which manages from roughly Montauk to the Hague Line separating U.S. and Canadian Exclusive Economic Zones.

10. NOAA Office of Sustainable Fisheries, "Federal Disaster Determination 56: New England Multispecies Groundfish, 2011–2013," *Fisheries Disaster Assistance*, http://www.nmfs.noaa.gov.

11. Joseph Chase Allen, "What of the Future of Fishing," in *Down to Sea for Fish: New Bedford Fishing Fleet* ([New Bedford, Mass.]: Reynold's, 1939), 22.

12. Allen, "What of the Future of Fishing," 22–23.

13. Daniel Pauly. "Anecdotes and the Shifting Baseline Syndrome of Fisheries," *Trends in Ecology and Evolution* 10, no. 10 (1995): 430.

INDEX

Acadian redfish (ocean perch). *See* redfish
Ackerman, Edward A., 92–94, 100–102
Against the Tide (Carey), 6
Albatross II (navy tug), 89
Alfaretta S. Snare (schooner), 42
Allen, Joseph Chase, 181–83
American Fisheries Society, 31, 79
American Fisheries Union (1880s), 36
American Fisheries Union (AFU): anticompetitive practices, 120; defense of union, 124; strike (1934), 103–8; strike (1946), 119; strike (1947), 119–20
American Fishery Union (Gloucester), 14–17
American values, 26–27. *See* "Maritime Revival"
Andrew, A. Piatt, 82–85, 87–91
antitrust: cod outfitters integration, 20; control of New England fish market, 41; General Seafoods Corporation purchase of Bay State Fishing Company, 107; Massachusetts investigation, 56; worker response, 23. *See also* Clayton Antitrust Act; Lapham bill
Atlantic Coast Fisheries: modernization, 121–22; "Taylor Method" innovation, 67. *See also* Cobe, Ira; Taylor, Harden

Atlantic Fisherman, 181
Atlantic Monthly, 45

Baird, Spencer, 25
Barker, B. Devereux, 67–68
Barnes, Clarence A., 119, 121, 124
Bartlett, Robert, 153
Bates, George, 108
Bates, William H., 148–50, 154, 161
Bay State Fishing Company, 49–50, 55, 65–66, 98–99, 107. *See also* Barker, B. Devereux
Bay State Halibut Company, 38
beam trawling: for halibut, 49–50; Nova Scotia bans, 52; reception, 50–55
Bell, Frank, 177–78
Bigelow, Henry Bryant, 76–77
Birdseye, Clarence, 68
Birdseye, Kellog, 75
Blackburn, Howard, 25–27, 187n23
Bland, Schuyler O., 87–88
Bluenose (schooner), 76
Boston: competition with Gloucester, 42, 51–52; fresh-fish fishing history, 37–38; market landscape, 38–39, 48–49; political activism, 11; T-Wharf fishing activity, 39–41, 51
Boston Fish Bureau, 49
Boston Fish Exchange, 100

201

Boston Fish Market Corporation (BFMC), 55–58, 59–60, 66, 190n51; cartel broken up, 59, 190n60
Boston Fish Pier: AFU strike, 103; consolidation of industry, 120; effect of the Depression, 86; halibut fishermen strike, 74; market crash, 117–19; mechanical processing, 68; medium and small draggers, 115; open auction, 55, 57, 70
Boston Food Fair, 103
Boston Fresh Fish Dealer's Association, 39
Boston Globe: Boston and Gloucester's competition, 42–43; on cartels, 38, 48–50, 59, 119; on fish market crash, 117; on foreign fleets, 179; on Gloucester's monopoly, 23; on the Magnuson Act, 1; on modernization, 123; on overfishing, 127, 140; on wartime profiteering, 62
Boston Journal, 40, 48, 51
Boy with the U.S. Fisheries, The (Rolt-Wheeler), 44
Bradford, Alfred, 49
Breedlove, Bernard, 99
Brown, James Scott, 73
Bulletin of the United States Fish Commission, 49
Bureau of Commercial Fisheries (BCF), 78; Fisheries Advisory Council, 107. *See also* U.S. Bureau of Fisheries (USBF)
Burke, John J., 140, 155
Business Advisory Council, Department of Commerce, 107

Cafasso, William, 159, 163, 165
Canada: American fisheries, competition with, 141; American fleet, search and seizure of, 15; fish sticks, 142; Nova Scotia and Newfoundland, processing plants, 143–44; tariff policy, 103; traditional mariners, 27
Captains Courageous (film), 91–93
Captains Courageous (Kipling), 9, 42, 44–45, 125

Carey, Richard Adams, 6
Carson, Rachel, 111, 114
cartels: cod outfitters integration, 20; control of New England fish market, 41; workers' response, 23. *See also* Boston Fish Market Corporation; Gigantic Fish Trust; Lapham bill
Century Illustrated Monthly, 15, 17
Chamberlayne, Charles Francis, 30–32, 33–34, 187n36
Champlain, Samuel de, 109
Chapman, W. M., 131, 135
Clayton Antitrust Act, 59, 119
Cleveland, Grover, 15–16
Cobe, Ira, 67
cod fishing: in Gloucester, 13; share system, 19; steam trawlers, 65; stocks, 127; transformation, 34. *See also* Salt Bounty; Treaty of Washington
cod outfitters: capital-intensive gear, 28; integration, 20, 28, 34
Cody, Joseph M., 154
Coffin, Frank M., 162
Colliers Weekly, 126
Collins, J. W., 15–17, 20, 24–26, 27, 31, 34, 49, 110
Columbia (dragger), 126–27
Conland, James, 45
Connolly, James Brendan, 53
conservation: benthic habitat, 78, 101, 146–47, 171, 177; fish stocks, sustainability of, 92–93, 178; international fisheries, regulation of, 134–35; juvenile haddock, 139, 172–73; modern gear, views of, 101; new species, strategy of harvesting, 111–12; overfishing, 126–27, 180; overfishing of haddock and redfish, 146; slashing quotas, 2; traditional fishing operations, 179–80; trophic cascade, 144–45; unity of labor and management forces, 130–32
Conservation Law Foundation (CLF), 180
Coplan, Maxwell Frederic, 109–10
Corliss, Beatrice, 158, 165
Crested Seas (Connolly), 53

INDEX 203

"Crisis in the Haddock Fishery, A" (Herrington), 111
Cunningham, Sylvester, 23

Daniel T. Church Company, 29
Davies, Joseph E., 107
Davies, Marjorie, 107
Davis Brothers, 140
Deep Sea Assembly of the Knights of Labor, 23, 41
Deep Sea's Toll, The (Connolly), 53
Dennis, William H., 61–62
Depression, effect of: decrease in fishing days, 97; fewer otter trawling ships, 96; small-scale fishermen's strategies, 96–97
"Destruction of Undersized Haddock on Georges Bank, 1947–1951" (Premetz), 173
"Developing Viewpoint of Oceanography, A" (Bigelow), 77
Difani, George, 154
Dingell, John, 163
Dobbs, David, 6
Doherty, Richard M., 144, 155
Donald, Gordon, 124–25
dory fishing, 25, 63, 98
Down to the Sea for Fish, 181. *See* Allen, Joseph Chase
draggers, 70, 75, 111, 115; for flounder, 112; racing, 76
Draheim, George P., 144, 155
Duff, James, 154
Dyer, Leslie, 158

East Coast Fisheries Company, 66
Economic Geography, 92
Economics of the New England Fishing Industry, The (Bell), 177
Edmonds, George, 85
Eisenhower, Dwight D., 141, 144, 156, 162, 165
Ericson, Thorwald, 109
Esperanto (schooner), 62
ethnicity: Anglo-Saxon fishermen, 33, 46–48, 63; fishermen demographics, 22; foreign labor, 24, 47, 122–23, 125; Native Americans, 46; Portuguese fishermen, 46, 86, 109, 123; xenophobia and nativism, 46–48, 63–64, 122–23

Federated Fishing Boats of New England and New York, 90
Federated Fishing Vessels of New England and New York, 130
Fiedler, Reginald Hobson, 69
Fish, Carl Russell, 72
Fish, Hamilton, 14
Fish and Fisheries Industries of the United States (Goode), 72
Fisheries Assistance Act of 1958, 160
fisherman, image of: in Bates bill, 149–50; bucolic perception, 24; in Congressional hearings (1956), 153; in Congressional hearings (1958), 158; in Congressional hearings (1959), 161–64, 167; cultural perception in New England, 7–8, 62–64, 71, 73–74, 79, 81, 100, 181; film version of New England fisherman, 92; in Gardner bill, 53–55; historical and traditional, 3–5, 7–8; historical ignorance, 183; in Lapham bill, 32–35; in loan guarantees, 87, 137–38; in outport fishing, 24–26, 28. *See also* newspapers
fishery disputes, international, 72–73
Fishing Gazette, 80, 90
fishing industry: antebellum, 18–20; Cold War opens market to European allies, 141, 156; complexities, 6; depleted fish stocks, 140–41, 147–49; fish sticks, impact of, 142–43; foreign fleets, 168, 170, 172, 174, 178–79; industrialization, 4; market crash, 65–66; political influence, loss of, 166–67; post–World War II, 115–19; small owner-operators pushed out, 20–22; ultimate fate, 181–83
fish sticks, 141–42
"Fish Trust Men Have Best Island Affords," 59–60
flounder: overfishing, 93, 101, 127, 144; trawling, 50, 115
flounder boats. *See* draggers
Food from the Sea (Carson), 114

204 INDEX

Fore River Ship and Engine Company, 50
Forest and Stream, 25
Four Fish (Greenberg), 6
Francis, Clarence, 107
free trade, 16, 41
fresh-fish fishing, 37
Fulham, Thomas, 143, 145
F/V *Harvard*, 107
F/V *Medford*, 105–6

Garcelon, William F., 52–55
Gardner, Augustus, 52–55
Gardner bill, 52–53. *See also* beam trawling; otter trawling
General Foods Corporation, 68
General Seafoods Corporation: Boston operations end, 143; Canadian operations, 106–8; racing sponsorship, 75; strike response, 103; technical innovations, 68
Gertrude L. Thebaud, Inc., 76
Gigantic Fish Trust, 40, 48, 50
Gilligan, Edmund, 126
Gloucester Board of Trade, 14, 31, 34
Gloucester Chamber of Commerce, 158
Gloucester Fisheries Association, 158
Gloucester Fisheries Commission, 154, 158, 163
Gloucester fishing industry: cartel control of halibut market, 39–40; criticism for failing to modernize, 109–10; fisheries politics, 178–79; Lapham bill, small-scale fishermen fight, 28–29; *List of Vessels Belonging to the District of Gloucester*, 21, 186n12; market power and profitability, 18, 24, 60; medium and small draggers revival, 111; political activism, 11; redfish, 111; small owner-operators pushed out, 20–22; stock depletion, 168; Viking grave claim, 109. *See also* races
Gloucester Vessel Owners Association, 158
"Good-By to the Dory Trotters" (McKaye), 121
Goode, George Brown, 71, 72

Gorton-Pew Fisheries Company, 22, 51, 59, 66, 85, 140, 142–43, 155; Canadian operations, 106, 108
Gould, Edwin W., 31–34, 63
Grace L. Fears (schooner), 25
Graham, John, 90
Graham, Michael, 77, 172
Grasso, Glenn M., 26
Great Gulf, The (Dobbs), 6
Green, Theodore F., 130–31, 133
Greenberg, Paul, 6
groundfish fishery, definition of, 6

haddock: Boston fleet, 6; "Current Haddock Situation on Georges Bank" (Schuck), 139; "Destruction of Baby Haddock on Georges Bank" (Schuck), 139; fishermen's strike, 50; heavy trawlers, destructiveness of, 78–79; overfishing worry, 76, 80, 89–90; "Protecting Baby Scrod Raises Production" (Schuck), 139; qualities for industrial processing, 68; scrod haddock (juvenile), 111, 128; steam trawler impact, 65. *See also* conservation
"Haddock and Cod Fishery Need a Research Vessel, The" (Graham), 90
halibut, 40, 49; for Boston fishermen, 47, 74; as a bottom-dweller, 49; in Gloucester, 100, 110
Halifax Herald, 61
Hand, Millet, 150
Hemeon, Bert, 126–27
Herrington, William C., 77–81, 89–90, 94, 111, 128, 133
Higgins, Elmer, 89–90
Higgins, James, 124–25
History of the New England Fisheries, A (McFarland), 72–73
Hoar, Frisbie, 23
Hodges, Wetmore, 76
Holbrook, Smith & Company, 37
Homer, Winslow, 9, 26, 187n23
Hopkinson, L. T., 131–32
House of Representatives, bills: defining "American Fishery," 108; extending agricultural relief to fishermen (H.R.

INDEX   205

9015), 87, 91; funding a fisheries research vessel (H.R. 8930), 89–90, 177; regional stock depletion treatment in hearings, 151–52, 154–55, 159, 164; State Department control to national fisheries commission, 152–56

House of Representatives, committees, 10; Committee on Merchant Marine and Fisheries, 30, 52–54, 108, 161; Committee on Merchant Marine, Radio, and Fisheries, 85–86

Huntsman, A. G., 77

I. J. Merritt (schooner), 43

industrialization: criticism of fisheries' big business, 121–22; criticism of non-modernized ports, 109; criticism of unionization, 122; "full-fledged" industry, 106; image of industry, 98–99, 103

International Coalition for the Exploration of the Seas (ICES), 76–77

International Convention on Northwest Atlantic Fisheries (ICNAF), 129–32; federal implementing legislation, 132–36; Gloucester fishing leaders argue for control, 138; Pacific fisheries, 135; romantic image argument for industry control, 137–38; self-regulation vs. federal control, 135–36

International Pacific Halibut Commission, 77

Jefferson, Thomas, 25
John Pew & Sons, 22
Joseph P. Johnson (schooner), 42

Kennedy, John F., 145, 159
Kipling, Rudyard, 42, 44–45, 125
Kuchel, Thomas, 152
Kurlansky, Mark, 6

labor unions: ambiguity of AFU, 119–20; Gloucester fleet workers' nationalities, 22–23; haddock fishermen's strike, 50; industrial uncertainty strikes, 141; New Bedford chapter of AFU supports strike, 104; New England Fish Exchange strike, 41; Seafood Workers Union (Gloucester) strike, 140; share system, 19, 58, 124–25; shutdown of trawler fleet, 143; standoff between processors and AFU, 119–20; strikes and lockouts, 47, 103–6; unionized fisheries workforce, 105. *See also* American Fisheries Union (AFU); Deep Sea Assembly of the Knights of Labor

Lane, Thomas, 149–50
Langford, Joseph, 75
Lapham, Oscar, 30
Lapham bill (1891), 28–34
Last Fish Tale, The (Kurlansky), 6

legislation, federal: bill to subsidize groundfish industry (1958), 156–58; bill to subsidize vessel construction (1959), 160–65; groundfish industry bill, national security argument, 159. *See also* antitrust; Fisheries Assistance Act of 1958; Magnuson Fisheries Conservation and Management Act; Merchant Marine Act of 1920; National Environmental Policy Act; Saltonstall-Kennedy Act

Lehlbach, Frederick, 85
Leonard, Richardson, 49

lobbying: control of federal policies, 152–56; Gloucester's role, 158–60, 165–66; heavy trawler fleet seeks subsidies, 145; reaction to ICNAF, 138; speaking for fishermen, 7. *See also* American Fishery Union; Gloucester Chamber of Commerce; Gloucester Fisheries Association; Gloucester Fisheries Commission; Gloucester Vessel Owners Association; New England Council; New England Fisheries Committee; U.S. Menhaden Oil and Guano Association

Lynch, Edward J., 144, 155, 178

MacDonald, Ishbel, 83
MacDonald, Ramsey, 82
MacInnis, William J., 88

mackerel fishing: Canadian grounds, 127; Gloucester, 46; Treaty of Washington, 13–14
MacManus, Tom, 61
Magnuson, Warren, 152, 154–55
Magnuson Fisheries Conservation and Management Act (MSA), 1, 3, 179–80
Maine Commission of Sea and Shore Fisheries, 32
Maine State Commissioner of Fisheries, 28
Manchester, Arthur, 29–30
Manchester v. Massachusetts, 30–31
Maritime History of Massachusetts (Morison), 73
"Maritime Revival," 26–27
Mary F. Chisholm (beam trawler), 49–50
Massachusetts Commissioners of Fish and Game, 22
Massachusetts Federation of Labor, 143
Massachusetts Fish Exchange, 38
Massachusetts General Court, committees on wartime profiteering and price fixing, 56–58
Massachusetts's Division of Marine Fisheries, 132
Matland, Randolph, 134
Mayflower (schooner), 75
McCormack, John W., 86–88, 90–91, 161
McFarland, Raymond, 71–74
McHugh, Patrick J., 104, 106, 108, 118–19, 120, 124, 130–36, 137–38, 145, 147
McKay, Ian, 7, 27
McKaye, Milton, 121–23
McKernan, Donald, 159
mechanized fishing: criticism of, 36; distrust by New Englanders, 60; effect on local stocks, 66, 126; image of traditional fishing, 94; market crash, 118; praise for New England, 122–23; waste from discarded fish, 78
Mellow, Joe, 76
Memorial Relating to the Destruction of State Fisheries, A (Gould), 32–33

menhaden fishery: industry consolidation, 29. *See also* U.S. Menhaden Oil and Guano Association
Merchant Marine Act of 1920 (Jones Act), 137–38
Metro-Goldwyn-Mayer, 91
Miller, George, 163
Morison, Samuel Eliot, 71, 73–74, 86, 149
Morrissey, Clayton, 82
Mudge, Raymond C., 69–70

Nash, Roderick, 8
National Environmental Policy Act, 3
National Fisheries Company, 49
National Geographic, 62, 81
National Marine Fisheries Service (NMFS), 1–3; groundfish disaster declarations, 181
nets: mesh size, 79–81, 94, 133, 139; "trouser" trawls, 80–81; Vigneron-Dahl trawling, 69–70, 78. *See also* otter trawling
New Bedford, 181
New England Committee for Aid to the Groundfish Industry, 156–58
New England Council, 158
New England Fish and Halibut Company, 50
New England Fisheries Committee, 146–47, 149, 153
New England Fishery Management Council, 180
New England Fish Exchange, 40–41, 55, 70, 93, 190n51
New England Fishing Company, 50–51
New England Fishing Economy, The (Doeringer, Moss & Terkla), 5
New England Magazine, 47
New England's Fishing Industry (Ackerman), 100
New Republic, 124
newspapers: antitrust investigation, 56, 59–60; hazards of fishing, 42–43; human interest in fisheries, 43
New York Times, 39, 92
New York Tribune, 53
Northeast Fisheries Science Center, 3
Northwest Atlantic Fisheries Conven-

tion. *See* International Convention on Northwest Atlantic Fisheries
Nugent, Frank S., 92

Oceanography (Bigelow), 77
O'Hara Brothers, 75, 140
Open Water (Connolly), 53
Ormond, Leonee, 45
otter trawling: catch increased, 65; declining productivity, 170; effect on haddock stocks, 114–15; fleet increase, 78; habitat destruction, 78, 192–93n39; haddock from Boston fleet, 55; mechanized fishing, 123; Nova Scotia bans, 52; portrayal of modernity, 99; Vigneron-Dahl trawl, 69
"Outlook of the Fisheries, The" (Collins), 15–16
Out of Gloucester (Connolly), 53

Palmer, George H., 31
Parkman, Francis, 71–73
"Passing of the New England Fisherman, The" (Thompson), 47
Pauly, Daniel, 183
Payne, Frederick, 152, 158
Philip P. Manta (schooner), 75
Pine, Ben, 75–76, 81–83, 126–27
political activism, 11
Poole, Gardner, 107
Premetz, Ernest, 139
price collapse, market crash, 65–67, 118, 125
public hearings, importance of, 83–84
public relations, 45; cod fisheries case, 15, 17; General Seafoods' strike response, 103–4; outport fishing images, 24–26, 28, 45; USMOGA and Lapham bill, 30. *See also* fisherman, image of; newspapers
Puritan (schooner), 75
purse seines, banned, 29

races: Gloucester revival, 75–76; international schooner, 61–62, 191n1
redfish: collapsing stock, 139–40; overfishing worry, 93, 128; replacement for haddock, 110–11, 114–15; trawl "revolution," 123
regulations, industry, 102
Resolute (beam trawler), 49–50
Rice, Thomas D., 118, 130–33, 137–38, 145–46, 153, 156–58, 162, 165
Rolt-Wheeler, Francis, 44
Roosevelt, Franklin Delano, 82, 91
Roscoe, Gerald, 117, 123
Rudder, The, 76
Ruskin, John, 148, 151
Russell, E. S., 90, 172
Russell, William, 33

Sabine, Lorenzo, 25
Salt Bounty (1789), 18–20, 23
salt cod fishing: consolidation and transformation, 34, 60; Gloucester, 13; impact on cod, 123; share system, 19. *See also* Salt Bounty; Treaty of Washington
Saltonstall, Leveret, 145
Saltonstall-Kennedy Act, 145–46, 149–51
Sandler, Solomon, 154–55, 162
Santos, Wayne, 61
Sargent, Francis W., 132, 147, 150
Saturday Evening Post, 99–100, 109, 121
scallops, 78, 113, 146
Schuck, Howard A., 139
Seafood Workers Union, 140
Seiners, The (Connolly), 53
Senate, "Extending Construction Loan Fund Benefits to Fishing and Whaling Industries" resolution, 82, 85–86, 88, 90
Shamrock (schooner), 75
small-scale fishermen: fighting Lapham bill, 28–29; opportunities for industry equality, 70–71; owner-operators pushed out, 20–22
Smith, Sylvanus, 14
Something of Myself (Kipling), 45
Soviet Bloc factory trawlers, 177–78
Spray (beam trawler), 49, 51, 65
State Department: biological conservation, 131–32, 134. *See also* Treaty of Washington

states' rights: local control of inshore waters, 29–31, 179. *See also* Lapham bill

statistics, catch: draggers, 70; effort data, 77, 80; federal fisheries data, 8–9; fish mortality peaks, 174; flounder, 112–13, 128–29; Gloucester's self-contained industry, 110; haddock and cod, 65, 70, 173; haddock and redfish, landings per day absent (LPDA), 168–69; labor strike impact, 121; landings data, 80, 139, 168; local vs. import catch, 118; mackerel, 111; Nova Scotia haddock, 176; overfishing impact, 126–28; redfish, 110–11, 113, 128–29, 174–75, 177; "shifting baselines," 183; spawning results, 80; undersized haddock destruction, 139; waste, 78, 101, 112–14, 115

statistics, fleet: cost of longer trips, 128; draggers, 70; haddock fleet, 172; heavy trawler trips, 96; otter trawlers, 170–71; trawling, 50

steam trawlers, 49, 53, 65

Storer, Humphrey, 25

Studds, Gerry, 179

Sylph (beam trawler), 49

tariffs. *See* U.S. Tariff Commission

Taylor, Harden, 67

technology: processing, freezing, and distribution, 67–68; sophisticated fishing tools, 4; Vigneron-Dahl trawl, 69–70. *See also* mechanized fishing

territorial waters. *See* Magnuson Fisheries Conservation and Management Act

Terry, David D., 87

Thebaud, Louis, 76

Thebaud (schooner), 76, 81; "Sail on Washington," lobbying trip, 82–85, 88, 91, 150

Thompson, William F., 77, 90

Thompson, Winfield, 47–48

Time, "Labor: Monopoly Broken," 121

Tollefson, Thor, 149

trawlers, definitions, 9

Treaty of Washington (1872, 1887), 13–14, 16–18

Trilling, Hy, 164

tub-trawling, 43

Tupper, Stanley, 147–48, 150

Turpin, Victor, 159, 166

United States Investor, 65–67

U.S. Bureau of Fisheries (USBF): otter trawling, research on destructiveness of, 9, 54, 64–65, 78, 101–2; otter trawling catch, 69–70; overfishing worry, 76; popular press, 43–44

U.S. Census Bureau, 22

U.S. Commissioner of Fisheries, 96, 194n1

U.S. Fish and Wildlife Service, 128, 173

U.S. Fish Commission, 22, 30, 31, 71

U.S. Menhaden Oil and Guano Association (USMOGA), 28–34

U.S. Nation Museum (Smithsonian), 21

U.S. Tariff Commission, 144, 156, 162–63, 164–65, 170

U.S. Treasury Department, T.D. 49682 on tariffs for American firm imports, 107–8

Wallace, Frederick, 63–64, 71

Webster, Daniel, 88

Webster, John R., 80

Welch, Marty, 82

White, Donald J., 127–28, 130, 139

Wilcox, W. A., 25

Wilderness and the American Mind (Nash), 8

Wood, Frank H., 98–99

Woods Hole, 51

Woods Hole Oceanographic Institution, 77–78

World War II, 113–14

Wright, Miriam, 7

xenophobia. *See under* ethnicity

yellowtail flounder, overfishing of, 128–29

www.ingramcontent.com/pod-product-compliance
Lightning Source LLC
Chambersburg PA
CBHW030650230426
43665CB00011B/1035